BTEC National N Ⅲ

Mathematics
for
Technicians

BTEC National N III

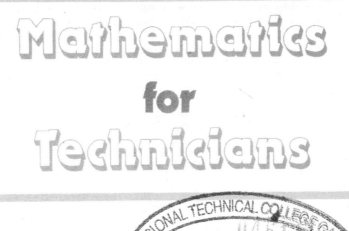

Mathematics for Technicians

A. Greer CEng, MRAeS
FORMERLY SENIOR LECTURER

G.W. Taylor BSc (Eng), CEng, MIMechE
FORMERLY PRINCIPAL LECTURER

Gloucestershire College
of Arts and Technology

STANLEY THORNES (PUBLISHERS) LTD

First published in 1989 by:
Stanley Thornes (Publishers) Ltd
Ellenborough House
Wellington Street
CHELTENHAM GL50 1YD
England

Reprinted 1990
Reprinted 1991
Reprinted 1992
Reprinted 1993
Reprinted 1994

British Library Cataloguing in Publication Data

Greer, A. (Alex).
 Mathematics for technicians.—2nd ed.
 BTEC National N3
 1. Mathematics
 I. Title   II. Taylor, G. W. (Graham William)
 III. Series
 510

ISBN 0-85950-932-X

510
GRE

Typeset by
Tech-Set, Gateshead, Tyne and Wear
Printed and bound in Great Britain at
BPC Wheatons Ltd

Contents

Authors' Note

This volume covers all the objectives included in the BTEC National NIII 'double' unit, and is for students of all disciplines.

Suitable examples have been included to cover applications for engineering, building construction, and science technologies.

A Greer
G W Taylor

Natural Logarithms and Logarithmic Graphs

OBJECTIVES

1 Define a natural (or Napierian) logarithm.

2 Determine natural logarithms from tables and by calculator.

3 State the relationship between natural and common logarithms.

4 Use natural logarithms to evaluate expressions arising from technology.

5 Solve equations involving e^x and $\log_e x$.

6 Recognise logarithmic scales.

7 Select appropriate numbers of cycles for plotting a given set of values on log-log graphs.

8 Plot values on log-log paper and determine the law.

9 Use the law from the above to determine other coordinates.

10 Select an appropriate number of cycles on the log scale, and an appropriate linear scale, for plotting a given set of values.

11 Plot values on log-linear paper and determine the law.

12 Use the law determined above to find other coordinates.

Theory of Logarithms

If N is a number such that

$$\boxed{N = a^x}$$

we say that x is the **logarithm** of N to the base a. We write

$$\boxed{\log_a N = x}$$

It should be carefully noted that

$$\text{NUMBER} = \text{BASE}^{\text{LOGARITHM}}$$

For example

since	$100 = 10^2$	we may write	$\log_{10} 100 = 2$
and as	$4.48 = e^{1.5}$	we may write	$\log_e 4.48 = 1.5$

1

Rules for the Use of Logarithms

These rules are true for any chosen value of the base:

(a) The logarithm of two numbers multiplied together may be found by adding their individual logarithms:

$$\log xy = \log x + \log y$$

(b) The logarithm of two numbers divided may be found by subtracting their individual logarithms:

$$\log \frac{x}{y} = \log x - \log y$$

(c) The logarithm of a number raised to a power may be found by multiplying the power by the logarithm of the number:

$$\log x^n = n \log x$$

Logarithms to the Base 10

Logarithms to the base 10 are called **common logarithms** and stated as '$\log_{10}$' (or 'lg'). When using logarithmic tables to solve numerical problems, we prefer tables to this base as they are simpler to use than tables to any other base. Common logarithms are also used for scales on logarithmic graph paper and also for calculations on the measurement of sound.

Logarithms to the Base e

In higher mathematics all logarithms are taken to the base e, where e = 2.718 28. Logarithms to this base are often called **natural logarithms**. They are also called **Napierian** or **hyperbolic logarithms**.

Natural logarithms are stated as '$\log_e$' (or 'ln').

Choice of Base for Calculations

If a scientific calculator is used then it is just as easy to use logarithms to the base e. Some machines have keys for both $\log_e$ and $\log_{10}$ but on the more limited models only $\log_e$ is given.

The natural logarithm is found by using the $\log_e$ (or ln) key and the natural antilogarithm is found by using the e^x key.

The common logarithm is found using the $\log_{10}$ (or lg) key and the common antilogarithm is found using the 10^x key.

Natural Logarithmic Tables

In most books of mathematical tables there is a table of natural logarithms. Part of such a table is shown below:

Hyperbolic, Natural or Napierian Logarithms

4.5	1.5041	5063	5085	5107	5129	5151	5173	5195	5217	5239	2	4	7	9 11 13 15 18 20
4.6	1.5261	5282	5304	5326	5347	5369	5390	5412	5433	5454	2	4	6	9 11 13 15 17 19
4.7	1.5476	5497	5518	5539	5560	5581	5602	5623	5644	5665	2	4	6	8 11 13 15 17 19
4.8	1.5686	5707	5728	5748	5769	5790	5810	5831	5851	5872	2	4	6	8 10 12 14 16 19
4.9	1.5892	5913	5933	5953	5974	5994	6014	6034	6054	6074	2	4	6	8 10 12 14 16 18
5.0	1.6094	6114	6134	6154	6174	6194	6214	6233	6253	6273	2	4	6	8 10 12 14 16 18
5.1	1.6292	6312	6332	6351	6371	6390	6409	6429	6448	6467	2	4	6	8 10 12 14 16 18
5.2	1.6487	6506	6525	6544	6563	6582	6601	6620	6639	6658	2	4	6	8 10 11 13 15 17
5.3	1.6677	6696	6714	6734	6752	6771	6790	6808	6827	6845	2	4	6	7 9 11 13 15 17
5.4	1.6864	6882	6901	6919	6938	6956	6974	6993	7011	7029	2	4	5	7 9 11 13 15 17

Natural logarithms of 10^{+n}

n	1	2	3	4	5	6	7	8	9
$\log_e 10^n$	2.3026	4.6052	6.9078	9.2103	11.5129	13.8155	16.1181	18.4207	20.7233

The first column of a set of full tables gives the natural logarithms of numbers from 1.0 to 9.9 (the specific table above only gives numbers from 4.5 to 5.49), and the tables are read in the same way as ordinary log tables except that the characteristic is also given. Thus

$$\log_e 4.568 = 1.5191$$

When a natural logarithm of a number that lies outside the tabulated range is required the subsidiary table has to be used. The following examples show how this is done.

Example 1.1

To find $\log_e 483.4$

Now
$$483.4 = 4.834 \times 100 = 4.834 \times 10^2$$

$\therefore$
$$\log_e 483.4 = \log_e (4.834 \times 10^2)$$
$$= \log_e 4.834 + \log_e 10^2$$

From the main table
$$\log_e 4.834 = 1.5756$$

From the subsidiary table
$$\log_e 10^2 = 4.6052$$

Thus
$$\log_e 483.4 = 1.5756 + 4.6052 = 6.1808$$

Example 1.2

To find $\log_e 0.053\,61$

Now
$$0.053\,61 = \frac{5.361}{100} = \frac{5.361}{10^2}$$

$\therefore$
$$\log_e 0.053\,61 = \log_e \left(\frac{5.361}{10^2} \right)$$
$$= \log_e 5.361 - \log_e 10^2$$

From the main table
$$\log_e 5.361 = 1.6792$$

From the subsidiary table
$$\log_e 10^2 = 4.6052$$

Thus
$$\log_e 0.053\,61 = 1.6792 - 4.6052$$
$$= -2.9260$$

Conversion of Common Logarithms to Natural Logarithms

The use of tables of natural logarithms is rather tedious as the last two worked examples show. Their use may be avoided by finding the common logarithm and then converting to a natural logarithmic value.

Suppose we have a number, N, and wish to find the value of its natural logarithm, that is $\log_e N$.

We have: $$\log_e N = \frac{\log_{10} N}{0.4343}$$

or $$\log_e N = 2.3026 \times \log_{10} N$$

Hence we may find the natural logarithm of a number by first finding its common logarithm and then multiplying by 2.3026 (or dividing by 0.4343).

Calculations Involving the Exponential Functions e^x and $\log_e x$

Example 1.3

Evaluate $50(e^{2.16})$.

The sequence of operations on a calculator is

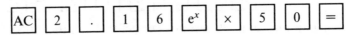

giving an answer 434 correct to 3 significant figures.

Example 1.4

Evaluate $200(e^{-1.34})$.

The sequence of operations would then be

giving an answer 52.4 correct to 3 significant figures.

Example 1.5

In a capacitive circuit the instantaneous voltage across the capacitor is given by $v = V(1 - e^{-t/CR})$ where V is the initial supply voltage, R

ohms the resistance, C farads the capacitance and t seconds the time from the instant of connecting the supply voltage.

If $V = 200$, $R = 10\,000$, and $C = 20 \times 10^{-6}$ find the time when the voltage v is 100 volts.

Substituting the given values in the equation we have

$$100 = 200(1 - e^{-t/(20 \times 10^{-6} \times 10\,000)})$$

$$\therefore \qquad \frac{100}{200} = 1 - e^{-t/0.2}$$

$$\therefore \qquad 0.5 = 1 - e^{-5t}$$

$$\therefore \qquad e^{-5t} = 1 - 0.5$$

$$\therefore \qquad e^{-5t} = 0.5$$

Thus in log form

$$\log_e 0.5 = -5t$$

$$\therefore \qquad t = -\frac{\log_e 0.5}{5}$$

The sequence of operations is

giving an answer 0.139 seconds correct to 3 significant figures.

Example 1.6

$$R = \frac{(0.42)S}{l} \times \log_e \left(\frac{d_2}{d_1} \right)$$

refers to the insulation resistance of a wire. Find the value of R when $S = 2000$, $l = 120$, $d_1 = 0.2$ and $d_2 = 0.3$

Substituting the given values gives

$$R = \frac{0.42 \times 2000}{120} \times \log_e \left(\frac{0.3}{0.2} \right)$$

$$= \frac{0.42 \times 2000}{120} \times \log_e 1.5$$

The sequence of operations would be

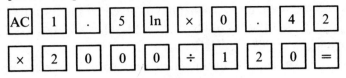

giving an answer 2.84 correct to 3 significant figures.

EXERCISE

1•1

1 Find the numbers whose natural logarithms are:
   (a) 2.76   (b) 0.677   (c) 0.09   (d) −3.46
   (e) −0.543   (f) −0.078

2 Find the values of:
   (a) $70e^{2.5}$ (b) $150e^{-1.34}$ (c) $3.4e^{-0.445}$

3 The formula

$$L = 0.000\ 644 \left(\log_e \frac{d}{r} + \frac{1}{4} \right)$$

is used for calculating the self-inductance of parallel conductors. Find L when $d = 50$ and $r = 0.25$.

4 The inductance (L microhenrys) of a straight aerial is given by the formula

$$L = \frac{1}{500} \left(\log_e \frac{4l}{d} - 1 \right)$$

where l is the length of the aerial in mm and d its diameter in mm. Calculate the inductance of an aerial 5000 mm long and 2 mm in diameter.

5 Find the value of $\log_e \left(\dfrac{c_1}{c_2} \right)^2$ when $c_1 = 4.7$ and $c_2 = 3.5$.

6 If $T = R \log_e \left(\dfrac{a}{a-b} \right)$ find T when $R = 28$, $a = 5$ and $b = 3$.

7 When a chain of length $2l$ is suspended from two points $2d$ apart on the same horizontal level

$$d = c \log_e \left(\frac{l + \sqrt{l^2 + c^2}}{c} \right)$$

If $c = 80$ and $l = 200$, find d.

8 The instantaneous value of the current when an inductive circuit is discharging is given by the formula $i = e^{-Rt/L}$. Find the value of this current, i, when $R = 30$, $L = 0.5$ and $t = 0.005$.

9 In a circuit in which a resistor is connected in series with a capacitor the instantaneous voltage across the capacitor is given by the formula $v = V(1 - e^{-t/CR})$. Find this voltage, v, when $V = 200$, $C = 40 \times 10^{-6}$, $R = 100\,000$ and $t = 1$.

10 In the formula $v = Ve^{-Rt/L}$ the values of v, V, R and L are 50, 150, 60 and 0.3 respectively. Find the corresponding value of t.

11 The instantaneous charge in a capacitive circuit is given by

$$q = Q(1 - e^{-t/CR})$$

Find the value of t when $q = 0.01$, $Q = 0.015$, $C = 0.0001$, and $R = 7000$.

Logarithmic Graphs

The Straight Line

Fig. **1.1**

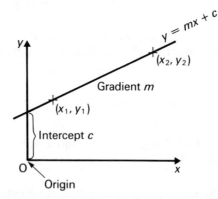

Remember the fundamental equation of the straight line:

$$y = mx + c$$

If m and c cannot be found directly from inspection of the graph, two points (x_1, y_1) and (x_2, y_2) on the line are chosen (Fig. 1.1) and their values substituted in the linear equation, giving the simultaneous equations:

$$y_1 = mx_1 + c$$
$$y_2 = mx_2 + c$$

from which m and c can be found.

Equations of the Type $z = at^n$, $z = ab^t$ and $z = ae^{bt}$

In all the work which follows in this chapter the logarithms used will be to the base 10 and are denoted by 'lg'.

Consider the following relationships in which z and t are the variables, whilst a, b and n are constants.

$z = at^n$

Fig. **1.2**

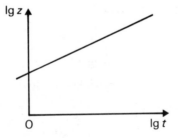

Now $\qquad z = at^n$

and taking logs $\qquad \lg z = \lg(at^n)$

$$= \lg t^n + \lg a$$

$$\lg z = n \lg t + \lg a$$

The given values of the variables will satisfy this equation if they satisfy the original equation. Comparing this equation with $y = mx + c$, which is the standard equation of a straight line, we see that if we plot $\lg z$ on the y-axis and $\lg t$ on the x-axis the result will be a straight line (Fig. 1.2) and the values of the constants n and a may be found using the two point method.

$z = ab^t$

Fig. **1.3**

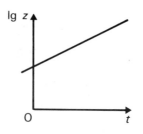

Now $\qquad z = ab^t$

and taking logs $\quad \lg z = \lg (ab^t)$

$$= \lg b^t + \lg a$$

$$\lg z = (\lg b)t + \lg a$$

We now proceed in a manner similar to that used for the previous equation by plotting $\lg z$ on the y-axis and t on the x-axis, and again obtain a straight line (Fig. 1.3).

$z = ae^{bt}$

Now $\qquad z = ae^{bt}$

and taking logs $\quad \lg z = \lg (ae^{bt})$

$$= \lg e^{bt} + \lg a$$

$$= (b \lg e)t + \lg a$$

but $\lg e = 0.4343$

$\therefore \qquad\qquad \lg z = (0.4343b)t + \lg a$

Again proceeding in a manner similar to that used for the previous equations, we plot $\lg z$ on the y-axis and t on the x-axis and again obtain a straight line (Fig. 1.3).

Logarithmic Scales

In your previous work you may have used ordinary graph paper for problems involving logarithmic and exponential laws. This entailed finding the logs of each individual given number on at least one of the axes. However if logarithmic scales are used it is no longer necessary to find individual logs.

We use logs to the base 10 since, as you will see, natural logs to the base e would result in inconvenient numbers on the scales.

It has been shown earlier that: Number = Base^{Logarithm}

and if we use a base of 10 then: Number = 10^{Logarithm}

Since	$1000 = 10^3$	then we may write	$\log_{10} 1000$	$= 3$
and since	$100 = 10^2$	then we may write	$\log_{10} 100$	$= 2$
and since	$10 = 10^1$	then we may write	$\log_{10} 10$	$= 1$
and since	$1 = 10^0$	then we may write	$\log_{10} 1$	$= 0$
and since	$0.1 = 10^{-1}$	then we may write	$\log_{10} 0.1$	$= -1$
and since	$0.01 = 10^{-2}$	then we may write	$\log_{10} 0.01$	$= -2$
and since	$0.001 = 10^{-3}$	then we may write	$\log_{10} 0.001$	$= -3$ and so on.

These logarithms may be shown on a scale as shown in Fig. 1.4.

Fig. **1.4**

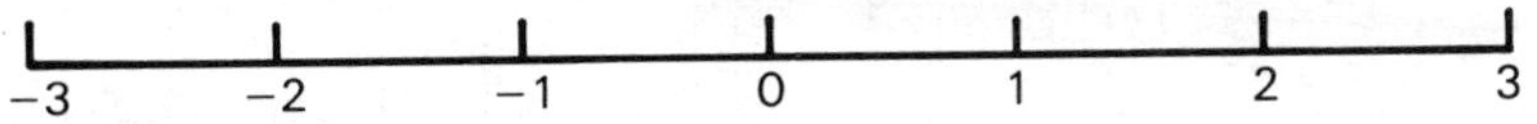

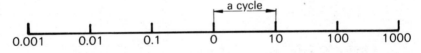

However, since we wish to plot numbers directly on to the scale (without any reference to their logarithms) the scale is labelled as shown in Fig. 1.5.

Fig. **1.5**

a cycle

0.001   0.01   0.1   0   10   100   1000

Each division is called a cycle and is sub-divided using a logarithmic scale as, for instance, the scales on a slide rule. Two such cycles are shown in Fig. 1.6.

Fig. **1.6**

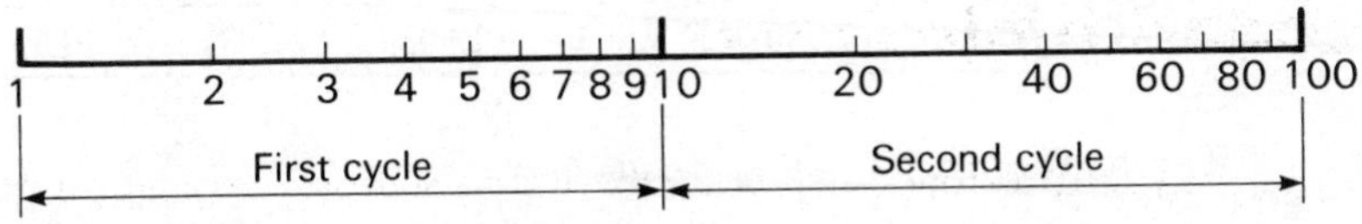

1   2   3   4   5 6 7 8 9 10   20   40   60 80 100

First cycle   Second cycle

The choice of numbers on the scale depends on the numbers allocated to the variable in the problem to be solved. Thus in Fig. 1.6 the numbers run from 1 to 100.

Logarithmic Graph Paper

Logarithmic scales may be used on graph paper in place of the more usual linear scales. By using graph paper ruled in this way log plots may be made without the necessity of looking up the logs of each given value. Semi-logarithmic graph paper is also available and has one way ruled with log scales whilst the other way has the usual linear scale. Examples of each are shown in Figs. 1.7 and 1.8.

Fig. **1.7**   Fig. **1.8**

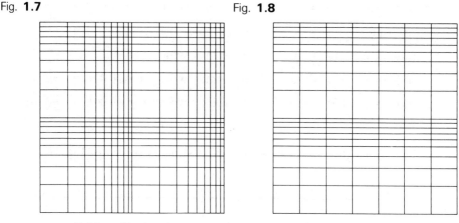

Full logarithmic rulings   Semi-logarithmic rulings

The use of these scales and the special graph paper is shown by the examples which follow.

Example 1.7

The law connecting two quantities z and t is of the form $z = at^n$. Find the law given the following pairs of values:

z	3.170	4.603	7.499	10.50	15.17
t	7.980	9.863	13.03	15.81	19.50

The relationship $z = at^n$

gives (see text) $\lg z = n \lg t + \lg a$ [1]

Hence we plot the given values of z and t on log scales as shown in Fig. 1.9.

Fig. **1.9**

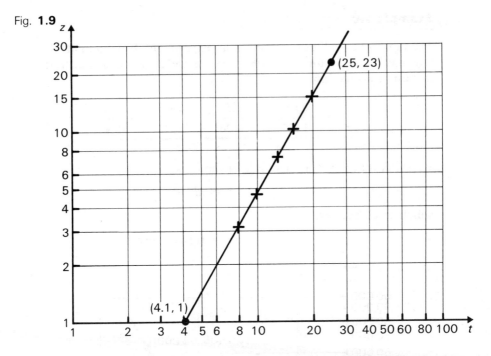

On both the vertical and horizontal axes we require 2 cycles, the first for values from 1 to 10 and the second for values from 10 to 100.

The constants are found by taking two pairs of coordinates:

Point $(25, 23)$ lies on the line, and putting these values in Equation [1],

$$\lg 23 = n \lg 25 + \lg a \qquad [2]$$

Point $(4.1, 1)$ lies on the line and putting these values in Equation [1],

$$\lg 1 = n \lg 4.1 + \lg a \qquad [3]$$

Subtracting Equation [3] from Equation [2],

$$\lg 23 - \lg 1 = n(\lg 25 - \lg 4.1)$$

$$\lg (23/1) = n\,[\lg (25/4.1)]$$

$$n = \frac{\lg(23/1)}{\lg(25/4.1)}$$

$$n = 1.73$$

Substituting this value of n in Equation [3],

$$\lg 1 = 1.73 \lg 4.1 + \lg a$$

$$\therefore \qquad \lg a = \lg 1 - 1.73 \lg 4.1$$

$$\therefore \qquad a = 0.087$$

Hence the law is $z = 0.087t^{1.73}$

Example 1.8

The table gives values obtained in an experiment. It is thought that the law may be of the form $z = ab^t$, where a and b are constants. Verify this and find the law.

t	0.190	0.250	0.300	0.400
z	11 220	18 620	26 920	61 660

We think that the relationship is of the form:

$$z = ab^t$$

which gives (see text),

$$\lg z = (\lg b)t + \lg a \qquad [1]$$

Fig. **1.10**

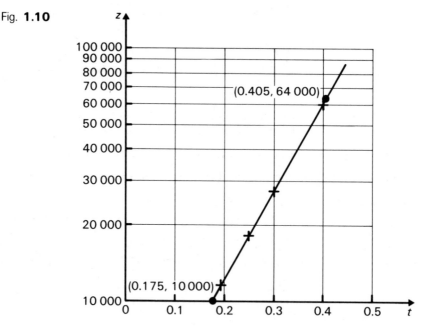

Hence we plot the given values of z on a vertical log scale — the t values, however, will be on the horizontal axis on an ordinary linear scale (Fig. 1.10).

The points lie on a straight line, and hence the given values of z and t obey the law. We now have to find the coordinates of two points lying on the line.

Point (0.405, 64 000) lies on the line, and substituting in Equation [1],

$$\lg 64\,000 = (\lg b)0.405 + \lg a \qquad [2]$$

Point (0.175, 10 000) lies on the line, and substituting in Equation [1],

$$\lg 10\,000 = (\lg b)0.175 + \lg a \tag{3}$$

Subtracting Equation [3] from Equation [2],

$$\lg 64\,000 - \lg 10\,000 = (\lg b)(0.405 - 0.175)$$

$$\therefore \qquad 4.8062 - 4.0000 = 0.230\,(\lg b)$$

$$\therefore \qquad \lg b = \frac{0.8062}{0.230} = 3.5052$$

$$\therefore \qquad b = 3200$$

Substituting in the Equation [3],

$$\lg 10\,000 = (3.5052)0.175 + \lg a$$

$$\therefore \qquad \lg a = \lg 10\,000 - 3.5052(0.175)$$

$$= 4 - 0.6134 = 3.3866$$

$$\therefore \qquad a = 2436$$

Hence the law is:

$$z = 2436(3200)^t$$

Example 1.9

V and t are connected by the law $V = ae^{bt}$. If the values given in the table satisfy the law, find the constants a and b.

t	0.05	0.95	2.05	2.95
V	20.70	24.49	30.27	36.06

The law is:

$$V = ae^{bt}$$

which gives (see text),

$$\lg V = 0.4343bt + \lg a \tag{1}$$

As in the last example V values are plotted on a log scale on the vertical axis, whilst the t values are plotted on the horizontal axis on an ordinary linear scale (Fig. 1.11).

Fig. **1.11**

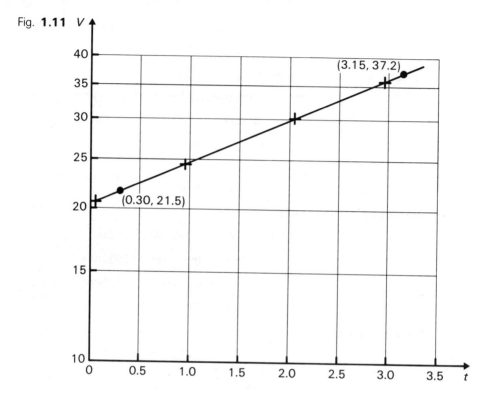

Point $(3.15, 37.2)$ lies on the line, and substituting in Equation [1],

$$\lg 37.2 = (0.4343)b(3.15) + \lg a \qquad [2]$$

Point $(0.30, 21.5)$ lies on the line, and substituting in Equation [1],

$$\lg 21.5 = (0.4343)b(0.30) + \lg a \qquad [3]$$

Subtracting,

$$\lg 37.2 - \lg 21.5 = (0.4343)b(3.15 - 0.30)$$

$$\therefore \qquad b = \frac{\lg(37.2/21.5)}{(0.4343)(2.85)}$$

Thus $\qquad b = 0.192$

Substituting in Equation [3]

$$\lg 21.5 = (0.4343)(0.192)(0.30) + \lg a$$

$$\therefore \qquad \lg a = 1.3324 - 0.0250$$

$$\therefore \qquad a = 20.3$$

EXERCISE

1·2

1 Using log-log graph paper show that the following set of values for x and y follows a law of the type $y = ax^n$. From the graph determine the values of a and n.

x	4	16	25	64	144	296
y	6	12	15	24	36	52

2 The following results were obtained in an experiment to find the relationship between the luminosity I of a metal filament lamp and the voltage V:

V	60	80	100	120	140
I	11	20	89	186	319

Allowing for the fact that an error was made in one of the readings show that the law between I and V is of the form $I = aV^n$ and find the probable correct value of the reading. Find the value of n.

3 Two quantities t and m are plotted on log-log graph paper, t being plotted vertically and m being plotted horizontally. The result is a straight line and from the graph it is found that:

when $\qquad m = 8, \quad t = 6.8$

and when $\qquad m = 20, \ t = 26.9$

Find the law connecting t and m.

4 The intensity of radiation R from certain radioactive materials at a particular time t is thought to follow the law $R = kt^n$. In an experiment to test this the following values were obtained:

R	58	43.5	26.5	14.5	10
t	1.5	2	3	5	7

Show that the assumption was correct and evaluate k and n.

5 The values given in the following table are thought to obey a law of the type $y = ab^{-x}$. Check this statement and find the values of the constants a and b.

x	0.1	0.2	0.4	0.6	1.0	1.5	2.0
y	175	158	60	32	6.4	1.28	0.213

6 The force F on the tight side of a driving belt was measured for different values of the angle of lap θ and the following results were obtained:

F	7.4	11.0	17.5	24.0	36.0
θ rad	$\pi/4$	$\pi/2$	$3\pi/4$	π	$5\pi/4$

Construct a graph to show these values conform approximately to an equation of the form $F = ke^{\mu\theta}$. Hence find the constants μ and k.

7 A capacitor and resistor are connected in series. The current i amperes after time t seconds is thought to be given by the equation $i = Ie^{-t/T}$ where I amperes is the initial charging current and T seconds is the time constant. Using the following values verify the relationship and find the values of the constants I and T:

i amperes	0.015 6	0.012 1	0.009 45	0.007 36	0.005 73
t seconds	0.05	0.10	0.15	0.20	0.25

8 For a constant pressure process on a certain gas the formula connecting the absolute temperature T and the specific entropy s is of the form $T = ke^{cs}$ where e is the logarithmic base and k and c are constants. When $T = 460$, $s = 1.000$, and when $T = 600$, $s = 1.089$. Find constants k and c to three significant figures.

9 The instantaneous e.m.f. v induced in a coil after a time t is given by $v = Ve^{-t/T}$, where V and T are constants. Find the values of V and T given the following values:

v	95	80	65	40	25
t	0.000 13	0.000 56	0.001 08	0.002 29	0.003 47

Polar Graphs

2

OBJECTIVES

1 Define polar coordinates.
2 State the relationship between polar and Cartesian coordinates.
3 Convert Cartesian to polar coordinates and vice versa.
4 Plot graphs of functions defined in polar coordinates such as $r = a$, $\theta = a$, $r = k\theta$, $r = \sin \theta$.

Polar Coordinates

You have met previously Cartesian (or rectangular) coordinates with which P may be given as the point (x, y) as shown in Fig. 2.1.

Fig. **2.1**   Fig. **2.2**

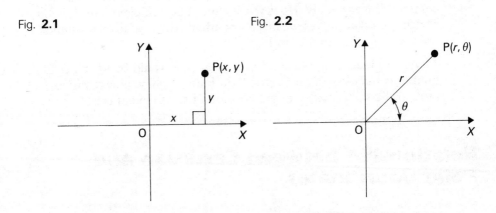

Another way of giving the location of P is by using its distance r from the origin O, together with angle θ that OP makes with the horizontal axis OX (Fig. 2.2). Figure 2.3 shows five typical points, plotted on a polar grid, together with their respective polar coordinates.

Fig. **2.3**

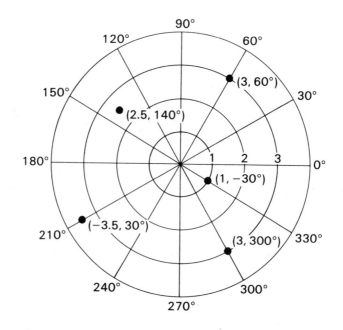

Positive values of θ are always measured anticlockwise from OX whilst negative values are measured clockwise (see Fig. 2.3): you will not often meet the latter case. A negative value of r means that the 'radius length' is extended 'backwards' through O from the normal angle position: see point $(-3.5, 30°)$ in Fig. 2.3. Note that in polar coordinates it is possible to define a point in more than one way: for example the point $(3, 300°)$ in Fig. 2.3 could also be defined as $(-3, 120°)$.

Polar graph paper is available but is not so readily obtained as the common linear variety. However, it should be sufficient here for you to sketch the polar curves we meet, or perhaps draw your own simple polar grid similar to that in Fig. 2.3.

A typical practical application of a polar plot would be to values of luminous intensity of an electric light bulb, to show how it varied around the bulb relative to the position of the filament etc.

Relationship between Cartesian and Polar Coordinates

From the right-angled triangle POM in Fig. 2.4 we can see that:

$$x = r \cos \theta$$

and $$y = r \sin \theta$$

Fig. **2.4**

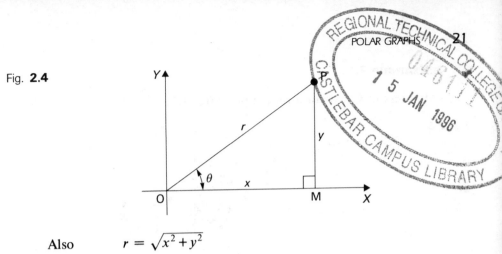

Also $\qquad r = \sqrt{x^2 + y^2}$

and $\qquad \tan\theta = \dfrac{y}{x}$

Using the above relationships it is reasonably easy to convert from Cartesian to polar coordinates, and vice versa. Always make a sketch of the problem because this will enable you to see which quadrant you are dealing with.

Example 2.1

Find the polar coordinates of the point $(-4, 3)$.

Fig. **2.5**

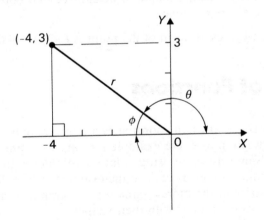

From Fig. 2.5 $\qquad \tan\phi = \dfrac{3}{4} = 0.75$

$\therefore \qquad\qquad\qquad \phi = 36.9°$

Thus $\qquad\qquad\qquad \theta = 180° - 36.9° = 143.1°$

Also $\qquad\qquad\qquad r = \sqrt{3^2 + 4^2} = 5$

Thus the point is $(5, 143.1°)$ in polar form.

Example 2.2

Express in Cartesian form the point $(6, 231°)$.

Fig. **2.6**

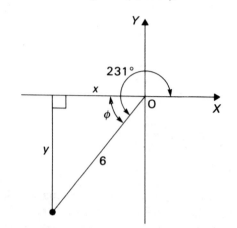

From Fig. 2.6 $\phi = 231° - 180° = 51°$

Thus $x = 6\cos 51° = 3.78$

and $y = 6\sin 51° = 4.66$

Now, having drawn a diagram, we can see that both the x and y coordinates are negative.

Thus the Cartesian form of the point is $(-3.78, -4.66)$.

Graphs of Functions

When using Cartesian coordinates a graph may be drawn to illustrate y as a function of x. For example $y = mx + c$ represents a straight line graph. Similarly when using polar coordinates a graph may be drawn to illustrate r in terms of θ. In the examples which follow we will sketch some of the more common polar graphs and this should enable us to become familiar with their shapes.

Example 2.3

Sketch the graph of $r = \sin \theta$ between $\theta = 0°$ and $\theta = 360°$.

From experience we know that $\sin \theta$ (and hence r) has a maximum value of $+1$, and a minimum value of -1. This will help initially in labelling our polar grid (Fig. 2.7). We may plot the graph from values obtained from our scientific calculator. We first measure off the angle θ and then measure off the length r along the angle boundary line.

Fig. **2.7**

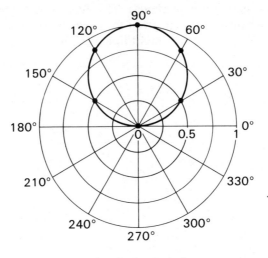

Graph of $r = \sin \theta$

Plotting from $\theta = 1$ to $\theta = 180°$ is straightforward. However, between 180° and 360° the values of r are negative and we merely repeat the curve already plotted. This curve is, in fact, a circle.

For interest you may like to verify that this is the only curve for *any* value of θ however large — both positive and negative.

Example 2.4

Sketch the graph of $r = 2$.

Here no mention is made of the angle θ. This means that $r = 2$ defines the graph whatever value is given for θ. Thus the graph is a circle of radius 2 as shown in Fig. 2.8.

Fig. **2.8**

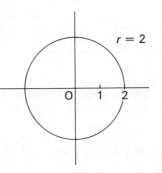

Example 2.5

Sketch the graph of $\theta = 40°$.

Here no mention is made of r. This means that $\theta = 40°$ defines the graph whatever value is given for r (whether positive or negative). Hence the graph is a straight line as shown in Fig. 2.9.

Fig. **2.9**

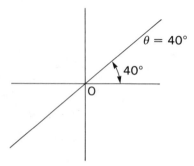

Example 2.6

Sketch the graph of $r = 3\theta$ for angle values equivalent to the range 0° to 540°.

As is usual in mathematics when an angle is used directly in calculations its value must be in radians. But we find it more convenient to plot the angle values using degrees and so we must be careful to find the corresponding angle values in radians when calculating r values.

Your scientific calculator may convert directly from degrees to radians. If not, we know that $360° = 2\pi$ rad or $1° = \dfrac{2\pi}{360}$ rad, and we may also use the constant multiplier facility to avoid entering this fraction for each r. A suitable sequence of operations would be:

For $\theta = 30°$:

$$\boxed{\text{AC}}\ \boxed{3}\ \boxed{\times}\ \boxed{2}\ \boxed{\times}\ \boxed{\pi}\ \boxed{\div}\ \boxed{360}\ \boxed{\times}\ \boxed{\times}\ \boxed{30}\ \boxed{=}$$

giving 1.57

and for other angles just enter its value,

e.g. for 60°:   giving 3.14

and for the highest value of 540°: $\boxed{540}\ \boxed{=}$ giving 28.27

The graph is known as an Archimedean spiral and is shown in Fig. 2.10.

Fig. **2.10**

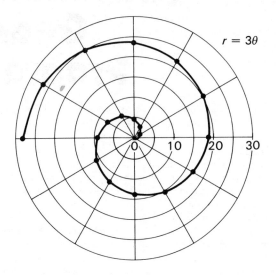

$r = 3\theta$

EXERCISE

1 Find the polar coordinates of the following points:

(a) $(5, 7)$ (b) $(-2, 3)$ (c) $(2, -3)$ (d) $(-3, -4)$

2 Find the Cartesian coordinates of the following points:

(a) $(2, 35°)$ (b) $(3, 127°)$ (c) $(1.5, 240°)$
(d) $(0.6, 312°)$ (e) $(2.3, -21°)$ (f) $(-5, 130°)$

Sketch the graphs for the polar equations in the questions which follow:

3 $r = 2\cos\theta$ 4 $r = 3$ 5 $\theta = 120°$

6 $r = \theta$ 7 $r = \sin^2\theta$ 8 $r = \cos^2\theta$

9 $r = 3\sin 2\theta$ 10 $r = \cos 3\theta$ 11 $r = a(1 + \cos\theta)$

Arithmetic and Geometric Progressions

OBJECTIVES

1 Express an A.P. in terms of *a* and *d*.

2 Determine an expression for the *n*th term of an A.P., and calculate the terms.

3 Calculate the parameters of an A.P. given any two pieces of relevant information.

4 Determine the expression for, and calculate the sum of, *n* terms of an A.P.

5 Express a G.P. in terms of *a* and *r*.

6 Determine the expression for the *n*th term of a G.P. and calculate terms.

7 Determine the expression for, and calculate the sum of, *n* terms of a G.P.

8 State the condition for convergence of a G.P.

9 Determine the expression for, and calculate the sum to infinity of, a convergent G.P.

10 Solve practical problems involving G.P.s, e.g. compound interest, range of speeds on a drilling machine.

Progression or Series

A set of numbers which are connected by a definite law is called a **series** or **progression**. Each of the numbers forming this set is called a **term** of the series. The following are examples of series:

$1, 3, 5, 7, \ldots$ (each term is obtained by adding 2 to the preceding one),
$1, 2, 4, 8, \ldots$ (each term is twice the preceding one),
$1, 4, 9, 16, 25, \ldots$ (the terms are the squares of successive integers).

The Arithmetic Progression

A series in which each term is obtained by adding or subtracting a constant quantity is called an **arithmetic progression** (or just **A.P.**). Thus the terms of the progression increase or decrease by a fixed amount which is called the **common difference**.

26

In the series

$$3, 6, 9, 12, 15, \ldots,$$

the difference between each term and the preceding one is 3, i.e. the common difference is 3. The series is therefore an A.P.

The series

$$10, 8, 6, 4, 2, 0, -2, -4, -6, \ldots$$

is an A.P. since each term is 2 less than the preceding one, i.e. the common difference is -2.

Some further examples of series in A.P. are:

$$0, 5, 10, 15, 20, \ldots \text{(common difference 5)},$$
$$1.2, 1.5, 1.8, 2.1, \ldots \text{(common difference 0.3)},$$
$$7.5, 6.1, 4.7, 3.3, \ldots \text{(common difference } -1.4).$$

General Expression for a Series in A.P.

Let the first number in the series be a and the common difference be d. The series can then be written as

$$a, \ a+d, \ a+2d, \ a+3d, \ldots$$

We can see that

the 1st term is	a
the 2nd term is	$a + (2-1)d$
the 3rd term is	$a + (3-1)d$
the 4th term is	$a + (4-1)d$, etc.

and hence any term of the series, say the nth term, is

$$\boxed{a + (n-1)d}$$

Example 3.1

Find the 9th and 18th terms of the series $1, 5, 9, 13, \ldots$

The 1st term of the series is 1, and so $a = 1$.

The common difference is 4, and so $d = 4$.

Hence the 9th term is

$$a + (9-1)d = 1 + (8 \times 4) = 33$$

and the 18th term is

$$a + (18-1)d = 1 + (17 \times 4) = 69$$

Example 3.2

The first term of a series in A.P. is 7 and the fifth term is 19. Find the eleventh term.

The 1st term $= a$

and the 5th term $= a + (5 - 1)d = a + 4d$

Hence $\qquad\qquad a = 7$

and $\qquad\qquad a + 4d = 19$

$\therefore\qquad\qquad\qquad 4d = 19 - a = 19 - 7 = 12$

$\therefore\qquad\qquad\qquad d = 3$

$\therefore\qquad\qquad$ 11th term $= a + (11 - 1)d$

$$= 7 + 10 \times 3$$

$$= 37$$

Example 3.3

Which term is 10.3 in the series $1.5, 2.6, 3.7, 4.8, \ldots$?

Here $a = 1.5$ and $d = 1.1$, and if we let the nth term $= 10.3$, then

$$a + (n - 1)d = 10.3$$

or $\qquad\qquad 1.5 + (n - 1)1.1 = 10.3$

$\therefore\qquad\qquad\qquad 1.1(n - 1) = 8.8$

$\therefore\qquad\qquad\qquad\qquad n = 9$

Hence 10.3 is the ninth term in the given series.

Example 3.4

Insert five arithmetic means between 2 and 26.

This question poses the problem of finding five numbers between 2 and 26 such that $2, ?, ?, ?, ?, ?, 26, \ldots$ form an A.P.

Now the 1st term is 2; thus

$$a = 2$$

and the 7th term is 26,

$\therefore$ $$a + 6d = 26$$

and hence from $a = 2$ we have

$$2 + 6d = 26$$

$\therefore$ $$d = 4$$

Hence the 2nd term $= a + d = 2 + 4 = 6$
and the 3rd term $= a + 2d = 2 + 8 = 10$
and the 4th term $= a + 3d = 2 + 12 = 14$
and similarly the other two terms are 18 and 22.

Cutting Speeds

The **cutting speed** for a centre lathe is the distance moved by a point on the surface of the work (such as point A in Fig. 3.1) in one minute. This is the speed at which the work moves past the tool point. It is also called the **surface speed** of the work. Cutting speeds are always quoted in metres per minute.

Fig. **3.1**

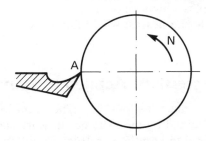

Let D (mm) $\quad$ = diameter of work,
$\quad$ N (rev/min) = spindle speed of lathe
and $\quad$ S (m/min) $\quad$ = cutting speed

Now the distance moved by point A over the surface in one rev is

$$\pi D \text{ mm} = \frac{\pi D}{1000} \text{ m}$$

Then the distance moved by point A over the surface in N revs is $\dfrac{\pi DN}{1000}$ metres.

Hence $\qquad$ cutting speed $\boxed{S = \dfrac{\pi DN}{1000}}$

This formula can be transposed to give

$$N = \frac{1000S}{\pi D} \quad \text{and} \quad D = \frac{1000S}{\pi N}$$

When milling or drilling the same formulae for cutting speeds etc. may be used, but then D would represent the diameter of the cutter or drill.

Speeds of Spindles and Feed Shafts; Approximations to Whole Numbers

These may be obtained exactly using formulae as above. However, the cutting speeds, although recommended by the manufacturers of materials and tools as the best values, are only approximate. Therefore in the calculations of spindle speeds etc. the figures obtained are rounded off to the nearest whole numbers. This is also the practice on the gearboxes of most machine tools where the whole numbers of teeth on gears also limit accuracy.

In the examples which follow in this chapter the exact values will be rounded off to the nearest whole numbers.

Arithmetic Progressions Applied to Spindle Speeds

In designing a machine tool the highest and lowest spindle speeds are usually determined by the extremes in work sizes which the machine will be required to handle. For example, a lathe may be designed to take a range of work varying from 25 mm to 375 mm diameter.

If we allow for a cutting speed of 20 m/min then the highest spindle speed, using the formula found in the last section, is

$$N = \frac{1000 \times 20}{\pi \times 25} = 255 \text{ rev/min}$$

to the nearest whole number which will be suitable for 25 mm diameter work.

The lowest spindle speed, suitable for 375 mm diameter work, is

$$N = \frac{1000 \times 20}{\pi \times 375} = 17 \text{ rev/min}$$

to the nearest whole number.

Suppose that the lathe has to have eight spindle speeds. Six intermediate spindle speeds must be chosen so that we have a series. One way of doing this is to choose the speeds so that they form a series in A.P.

In the series, the first term

$$a = 17$$

and the eighth term

$$a + 7d = 255$$

$$\therefore \qquad 17 + 7d = 255$$

$$\therefore \qquad d = 34$$

Hence all the speeds can be found, the second speed being

$$a + d = 17 + 34 = 51$$

and so on, giving the eight speeds as follows:

$$17, 51, 85, 119, 153, 187, 221 \text{ and } 255 \text{ rev/min}$$

Example 3.5

A lathe is to have six spindle speeds arranged in A.P. It is to cater for work ranging from 12 mm to 250 mm diameter. Allowing for a cutting speed of 25 m/min, find these spindle speeds.

Since $N = \dfrac{1000S}{\pi D}$, the first speed $= \dfrac{1000 \times 25}{\pi \times 250} = 32$ rev/min

and the sixth speed $= \dfrac{1000 \times 25}{\pi \times 12} = 663$ rev/min

and since the speeds are in A.P.,

$$a = 32$$

and $$a + 5d = 663$$

$$\therefore \qquad 32 + 5d = 663$$

$$\therefore \qquad d = 126.2$$

Hence the second speed

$$a + d = 32 + 126.2 = 158 \text{ rev/min}$$

the third speed

$$a + 2d = 32 + 2(126.2) = 284 \text{ rev/min}$$

and so on, giving the six spindle speeds (to the nearest whole numbers) as follows:

$$32, 158, 284, 411, 537 \text{ and } 663 \text{ rev/min}$$

Sum of a Series in Arithmetic Progression

Consider the typical A.P. series already met, supposing it to consist of n terms of which the last term is l.

The 1st term is $\qquad a$
the 2nd term is $\qquad a + d$
the 3rd term is $\qquad a + 2d$, etc.
. . . .
and the last term is $l = a + (n - 1)d$

Now if we write down the sum S of the series starting with the first term a, we have

$$S = a + (a + d) + (a + 2d) + \ldots + l$$

and if we write down the sum S of the series starting with the last term l, we have

$$S = l + (l - d) + (l - 2d) + \ldots + a$$

If we add these two equations we get

$$2S = (a + l) + (a + l) + (a + l) + \ldots + (a + l)$$

since all the d terms cancel out. Since there are n terms there will be n lots of $(a + l)$;

hence

$$2S = n(a + l)$$

or

$$\boxed{S = \frac{n}{2}(a + l)}$$
$$\text{where} \quad l = a + (n - 1)d$$

Example 3.6

Find the sum of the series $1.5, 2.6, 3.7, 4.8, \ldots$, if there are altogether 12 terms.

The last term

$$l = a + (n-1)d$$

In this case $a = 1.5$, $n = 12$ and $d = 1.1$ (the difference between any two adjacent terms), and substituting these values in the expression for l we get

$$l = 1.5 + (12 - 1)1.1$$
$$= 1.5 + 12.1$$
$$= 13.6$$

But the sum

$$S = \frac{n}{2}(a + l)$$

and on substituting our values,

$$S = \frac{12}{2}(1.5 + 13.6)$$
$$= 6 \times 15.1$$
$$= 90.6$$

Example 3.7

Find the sum of the series $3, 1, -1, -3, \ldots, -45$.

We need first to find the number of terms n.

Now the last term

$$l = a + (n-1)d$$

We have $a = 3$, $l = -45$ and $d = -2$, and substituting these values we get

$$-45 = 3 + (n-1)(-2)$$
$$\therefore \qquad 2(n-1) = 48$$
$$\therefore \qquad n = 25$$

Now the sum

$$S = \frac{n}{2}(a + l)$$

and substituting our values,

$$S = \frac{25}{2}[3 + (-45)]$$
$$= (12.5)(-42)$$
$$= -525$$

EXERCISE

3·1

1 Find the 8th and 17th terms of the series 12, 15, 18, 21, . . .

2 Which term of the series −40, −38, −36, . . . is equal to 4?

3 The 3rd term of an A.P. is −8 and the 16th is 57. Find
(a) the 10th term,
(b) the sum of the first 14 terms.

4 Insert six arithmetic means between 3.30 and 6.45.

5 Three numbers are in A.P. The product of the 1st term and the last term is −140. The sum of three times the 2nd term and twice the 1st term is −14. Find the numbers.

6 How many terms in the series 5.6, 4.1, 2.6, . . . must be taken so that their sum is −3.6?

7 In boring a well 300 m deep, the cost of boring the first metre is £200 000 whilst each subsequent metre costs £250. What is the cost of boring the entire well?

8 Water fills a tank at a rate of 150 litres during the first hour, 350 litres during the second hour, 550 litres during the third hour and so on. Find the number of hours necessary to fill a rectangular tank 16 m × 7 m × 7 m.

9 A body falling freely falls 5 m in the first second of its motion, 15 m in the second second, 25 m during the third second and so on.
(a) How far does it fall in 12 s?
(b) How far does it fall in the 14th second?
(c) How long will it take to fall 3000 m?

10 A lathe has to have five spindle speeds arranged in A.P. It is to cater for work ranging from 25 mm to 200 mm in diameter. Allowing for a cutting speed of 25 m/min, find the spindle speeds.

The Geometric Progression

A series in which each term is obtained from the preceding term by multiplying or dividing by a constant quantity is called a **geometric progression** or simply a **G.P.** The constant quantity is called the **common ratio** of the series.

In the series $1, 2, 4, 8, 16 \ldots$, each successive term is formed by multiplying the preceding one by 2. The series is therefore in G.P., and the common ratio is 2.

The series $6, 2, 2/3, 2/9, \ldots$ is a G.P. since each successive term is formed from the preceding by multiplying by $\frac{1}{3}$, and this is also the common ratio.

Some further examples of series in G.P. are:

$$2, 10, 50, 250, \ldots \text{(common ratio 5),}$$
$$4, -8, 16, -32, \ldots \text{(common ratio } -2),$$
$$1.3, 1.95, 2.925, \ldots \text{(common ratio 1.5).}$$

A General Expression for a Series in G.P.

Let the first term be a and the common ratio r. The series can be represented by $a, ar, ar^2, ar^3, \ldots$

We see that: the 1st term is a,

the 2nd term is ar (the index of r is 1),

the 3rd term is ar^2 (the index of r is 2),

the 4th term is ar^3 (the index of r is 3), etc.

The index of r is always one less than the number of the term in the series.

Hence   the 9th term is ar^8

the 20th term is ar^{19}

and   the nth term is $\boxed{ar^{n-1}}$

Example 3.8

Find the seventh term of the series $2, 6, 18, \ldots$

The 1st term is 2 and the common ratio is 3, that is, $a = 2$ and $r = 3$.

Hence the 7th term is

$$ar^6 = 2 \times 3^6 = 1458$$

Example 3.9

The 1st term of a series in G.P. is 19, and the 6th is 27. Find the 10th term.

1st term

$$a = 19$$

$$\text{and the 6th term} = ar^5 = 27$$

$\therefore$ $$19r^5 = 27$$

$\therefore$ $$r = \sqrt[5]{\frac{27}{19}} = 1.073$$

Hence the 10th term $= ar^9 = 19 \times 1.073^9 = 35.8$

Example 3.10

Insert four geometric means between 1.8 and 11.2.

This is equivalent to putting four terms between 1.8 and 11.2 so that the six numbers form a series in G.P. Thus $1.8, ?, ?, ?, ?, 11.2$ must be a G.P.

The 1st term is 1.8.

$\therefore$ $$a = 1.8$$

The 6th term is 11.2.

$\therefore$ $$ar^5 = 11.2$$

that is, $$1.8r^5 = 11.2$$

$\therefore$ $$r = \sqrt[5]{\frac{11.2}{1.8}} = 1.441$$

Hence the 2nd term,

$$ar = 1.8 \times 1.441 = 2.59$$

the 3rd term,

$$ar^2 = 1.8 \times 1.441^2 = 3.74$$

the 4th term,

$$ar^3 = 1.8 \times 1.441^3 = 5.39$$

and the 5th term,

$$ar^4 = 1.8 \times 1.441^4 = 7.76$$

Hence the required geometric means are 2.59, 3.74, 5.39 and 7.76.

Geometric Progressions Applied to Spindle Speeds

Previously we have considered spindle speeds in A.P. They may also be arranged in G.P. and, as will be seen later, this is a better way of arranging them.

Example 3.11

A drill is to have seven speeds arranged in G.P. It is to drill a range of holes from 3 mm to 12 mm in diameter at a cutting speed of 15 m/min. Find the seven spindle speeds.

Since $N = \dfrac{1000S}{\pi D}$, the 1st speed $= \dfrac{1000 \times 15}{\pi \times 12}$

$$= 398 \text{ rev/min}$$

and the 7th speed is $\dfrac{1000 \times 15}{\pi \times 3} = 1592 \text{ rev/min}$

Since all the speeds are to be in G.P.,

$$a = 398$$

and $$ar^6 = 1592$$

$\therefore \qquad 398r^6 = 1592$

$\therefore \qquad r = \sqrt[6]{\dfrac{1592}{398}} = 1.26$

Hence the 2nd speed is

$$ar = 398 \times 1.26 = 501 \text{ rev/min}$$

and the 3rd speed is

$$ar^2 = 398 \times 1.26^2 = 632 \text{ rev/min}$$

which are rounded off to the nearest whole numbers giving the required speeds as

$$398, 501, 632, 796, 1003, 1264 \text{ and } 1592 \text{ rev/min}$$

Comparison of Spindle Speeds in A.P. and in G.P.

In this section we shall show that spindle speeds in G.P. are to be preferred to those in A.P. This will be done by considering the following example:

Example 3.12

A lathe tool has to accommodate work between 25 mm and 300 mm in diameter. Six spindle speeds are required and the cutting speed is to be 25 m/min. Find the six speeds, (a) if they are in A.P., (b) if they are in G.P. In both cases find the work diameter appropriate to each spindle speed.

$$\text{1st speed} = \frac{1000 \times 25}{\pi \times 300} = 27 \text{ rev/min}$$

$$\text{6th speed} = \frac{1000 \times 25}{\pi \times 25} = 318 \text{ rev/min}$$

(a) *Speeds in A.P.*

$$\text{1st speed } a = 27$$

$$\text{and 6th speed} = a + 5d = 318$$

$$\therefore \qquad 27 + 5d = 318$$

$$\therefore \qquad d = 58.2$$

$$\therefore \qquad \text{2nd speed} = a + d = 85 \text{ rev/min}$$

$$\text{and 3rd speed} = a + 2d = 143 \text{ rev/min}$$

Hence arranged in A.P. the spindle speeds are

$$27, 85, 143, 202, 260 \text{ and } 318 \text{ rev/min}$$

The bar diameters appropriate to each speed can be calculated by the formula

$$D = \frac{1000 \times S}{\pi \times N}$$

These are, in order, 295, 94, 56, 39, 31 and 25 mm.

(b) *Speeds in G.P.*

$$\text{1st speed } a = 27$$

$$\text{and 6th speed} = ar^5 = 318$$

$$\therefore \qquad 27r^5 = 318$$

$$\therefore \qquad r = \sqrt[5]{\frac{318}{27}} = 1.638$$

$$\therefore \qquad \text{2nd speed} = ar = 44$$

$$\text{and} \qquad \text{3rd speed} = ar^2 = 72$$

Hence, arranged in G.P., the spindle speeds are

$$27, 44, 72, 119, 194 \text{ and } 318 \text{ rev/min}$$

The corresponding bar diameters can be calculated as before and are, in order, 295, 181, 111, 67, 41 and 25 mm

For convenience the figures obtained for the speeds arranged in both A.P. and G.P. are tabulated below:

Speed number	A.P.		G.P.	
	Speed	Bar diam.	Speed	Bar diam.
1	27	295	27	295
2	85	94	44	181
3	143	56	72	111
4	202	39	119	67
5	260	31	194	41
6	318	25	318	25

The graph (Fig. 3.2) shows that when the speeds are in A.P. there are too many at the higher end of the speed range and too few at the lower end. If we wish, for instance, to turn a 175 mm diameter bar the correct spindle speed would be 45 rev/min to give a cutting speed of 25 m/min. We can either use speed number 1 (27 rev/min) or speed number 2 (85 rev/min). Speed number 1 is too low and speed number 2 is too high. Spindle speeds in G.P. give a far better distribution. For any diameter we can find a spindle speed which will give a near approximation to the correct cutting speed of 25 m/min. It is for this reason that the spindle speeds are usually arranged in G.P. rather than in A.P.

Fig. **3.2**

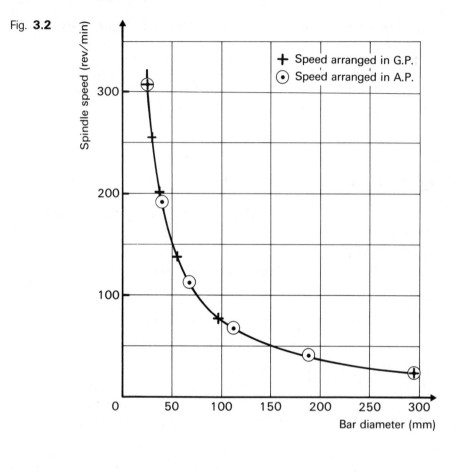

Fig. **3.3**

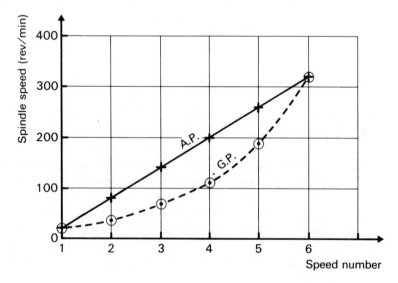

The graph (Fig. 3.3) shows spindle speed plotted against speed number. It can be seen that the graph is straight for A.P. and curved for G.P. This illustrates the better distribution of spindle speeds when they are arranged in G.P. The G.P. arrangement gives closer speed intervals at the lower end of the range and wider intervals at the higher end.

The Sum of a Series in Geometric Progression

Consider the typical G.P. series already met, supposing it to consist of n terms.

The 1st term is a,

the 2nd term is ar,

the 3rd term is ar^2

. . . .

and the last term is ar^{n-1}

Hence the sum S of the series is given by

$$S = a + ar + ar^2 + ar^3 + \ldots + ar^{n-1}$$

If we now multiply the above equation throughout by r we get

$$Sr = ar + ar^2 + ar^3 + \ldots + ar^{n-1} + ar^n$$

On subtracting this equation from the one for S we get

$$S - Sr = a - ar^n$$

that is, $$S(1 - r) = a(1 - r^n)$$

$\therefore$

$$S = \frac{a(1 - r^n)}{1 - r}$$

which is a convenient form to use if $r < 1$.

Now if we multiply both top and bottom of the right-hand side of this equation by -1 we have

$$S = \frac{a(r^n - 1)}{r - 1}$$

which is easier to use if $r > 1$.

The Sum to Infinity of a G.P.

Consider a G.P. series having first term $a = 1$ and common ratio $r = 2$.

Its sum would be $1 + 2 + 4 + 8 + 16 + 32 + 64 + \ldots$

We can see that the terms are getting successively larger. Thus for an infinite number of terms it is *not* possible to find a value for their sum.

However, for a G.P. series having $a = 1$ and $r = \dfrac{1}{3}$

the sum would be $1 + \dfrac{1}{3} + \dfrac{1}{9} + \dfrac{1}{27} + \dfrac{1}{81} + \dfrac{1}{243} + \dfrac{1}{729} + \ldots$

Here the terms are rapidly getting smaller and a numerical answer to any reasonable degree of accuracy may soon be obtained. Such a series is said to be **convergent**.

In general for a G.P. series having $r < 1$, then it is convergent. If we consider the expression for the sum $S = \dfrac{a(1 - r^n)}{1 - r}$ then for an infinite series (which means that n is infinitely large) the term r^n will be virtually zero. Thus the sum to infinity will be $\dfrac{a}{1 - r}$.

N.B. An A.P. series will always be non convergent, so it is *not* possible to find its sum to infinity.

Example 3.13

Find the sum of the series $2, 6, 18, \ldots$ which has 6 terms.

Since $r = 3$ (obtained from 6/2 or 18/6, etc.), that is $r > 1$, we shall use the form

$$S = \frac{a(r^n - 1)}{r - 1}$$

In this question $a = 2$ and $n = 6$. Substituting these values we get

$$S = \frac{2(3^6 - 1)}{3 - 1}$$

$$= \frac{2(729 - 1)}{2}$$

$$= 728$$

Example 3.14

Find the sum of the series $8, -4, 2, \ldots -\frac{1}{64}$.

Here $a = 8, r = -\frac{1}{2}$ (from $-4/8$ or $2/(-4)$, etc.), and if we assume the series has n terms then the last term is ar^{n-1}. However we are given this is $-\frac{1}{64}$.

$\therefore$
$$ar^{n-1} = -\frac{1}{64}$$

$\therefore$
$$8\left(-\frac{1}{2}\right)^{n-1} = -\frac{1}{64}$$

$\therefore$
$$\left(-\frac{1}{2}\right)^{n-1} = -\frac{1}{64 \times 8}$$

$\therefore$
$$\left(-\frac{1}{2}\right)^{n-1} = -\frac{1}{2^6 \times 2^3}$$

$\therefore$
$$\left(-\frac{1}{2}\right)^{n-1} = -\frac{1}{2^9}$$

from which, by inspection,

$$n - 1 = 9$$

$\therefore$
$$n = 10$$

If it was not possible to obtain $n - 1$ by inspection, the equation would have needed the use of logs for solution. This method is shown in a later example.

We now have $a = 8, r = -\frac{1}{2}$ and $n = 10$, and if we substitute these values in the expression

$$S = \frac{a(1 - r^n)}{1 - r} \quad \text{since } r < 1,$$

we have

$$S = \frac{[1 - (-\frac{1}{2})^{10}]}{1 - (-\frac{1}{2})}$$

$$= \frac{8(1 - 1/1024)}{1 + \frac{1}{2}}$$

$$= 5.328$$

Compound Interest

Consider a sum of £P, invested at r% for n years.

The initial value is £P, and after 1 year the value is

$$£\left(P + \frac{Pr}{100}\right) \quad \text{or} \quad £P\left(1 + \frac{r}{100}\right)$$

In the 2nd year the interest will be on

$$£P\left(1 + \frac{r}{100}\right)$$

so after the 2nd year the value is

$$£\left[P\left(1 + \frac{r}{100}\right) + P\left(1 + \frac{r}{100}\right) \times \frac{r}{100}\right]$$

$$= £\left[P\left(1 + \frac{r}{100}\right)\left(1 + \frac{r}{100}\right)\right]$$

$$= £P\left(1 + \frac{r}{100}\right)^2$$

Similarly after the 3rd year the value is

$$£P\left(1 + \frac{r}{100}\right)^3$$

and the amount after n years is obtained from the G.P. whose first term is £P and whose common ratio is $(1 + r/100)$, and equals

$$\boxed{£P\left(1 + \frac{r}{100}\right)^n}$$

Example 3.15

Calculate the value of £2500 invested at 5% compound interest after eight years.

Here $P = 2500$, $r = 5$ and $n = 8$, and substituting these values in the formula £$P(1 + r/100)^n$ we find the required value to be

$$£2500\left(1 + \frac{5}{100}\right)^8$$

$$= £2500(1.05)^8$$

$$= £3693.64$$

Example 3.16

In the first week of production 1000 articles are made. If the rise in weekly production is 5% per week, how many weeks are necessary to produce a total of 15 000 articles?

In the 1st week,

$$\text{production} = 1000$$

In the 2nd week,

$$\text{production} = 1000 + 1000 \left(\frac{5}{100} \right)$$

$$= 1000 \left(1 + \frac{5}{100} \right)$$

In the 3rd week,

$$\text{production} = 1000 \left(1 + \frac{5}{100} \right)^2$$

As in compound interest calculations the weekly production rates form a series:

$$1000, \quad 1000 \left(1 + \frac{5}{100} \right), \quad 1000 \left(1 + \frac{5}{100} \right)^2, \quad \text{etc.,}$$

that is,

$$1000, \quad 1000(1.05), \quad 1000(1.05)^2, \quad \text{etc.,}$$

which is a G.P. having $a = 1000$ and $r = 1.05$.

Now

$$S = \frac{a(r^n - 1)}{r - 1}$$

We have to find the number of weeks n for $S = 15\,000$

$$\therefore \qquad 15\,000 = \frac{1000(1.05^n - 1)}{1.05 - 1}$$

from which

$$1.05^n = 1.75$$

To solve this equation for n we take logs of both sides.

$$\log 1.05^n = \log 1.75$$

$\therefore$
$$n \times \log 1.05 = \log 1.75$$

$\therefore$
$$n = \frac{\log 1.75}{\log 1.05}$$

$\therefore$
$$n = 11.47 \text{ weeks.}$$

EXERCISE

3·2

1 Find the 8th term of the series, $1, 1.1, 1.21, \ldots$

2 Find the 8th term of the series which is a G.P. having a 2nd term of -3 and a 5th term of 81.

3 Insert 3 geometric means between $1\frac{1}{8}$ and $\frac{1}{72}$.

4 Find the sum of the series $60, 30, 15, \ldots$, (a) to 6 terms, (b) to infinity.

5 The sum of the first 6 terms of a G.P. series is 189 and the common ratio is 2. Find the series.

6 On a certain lathe the cone and back gear give 8 possible spindle speeds. If the least speed is 3 rev/min and the greatest speed is 300 rev/min, find the intermediate spindle speeds, if they are in G.P.

7 A lathe has 6 speeds arranged in G.P. It is to accommodate work from 12 mm diameter to 150 mm diameter at a cutting speed of 22 m/min. Find the six spindle speeds.

8 A drill has to drill holes between 2 mm and 25 mm in diameter. Eight spindle speeds are required and the cutting speed is to be 20 m/min. Find the eight speeds if (a) they are in A.P., (b) they are in G.P.

Illustrate your results by means of a graph, on which the spindle speeds (rev/min) are plotted vertically and bar diameters (mm) are plotted horizontally. Comment on the results, stating which system is preferable and why.

9 A contractor hires out machinery. In the first year of hiring out one piece of equipment the profit is £6000, but this diminishes by 5% on successive years. Show that the annual profits form a G.P. and find the total of all profits for (a) the first 6 years, (b) all possible future years.

10 A firm starts work with 110 employees for the 1st week. The number of employees rises by 6% per week.

   (a) How many weeks will it take for 250 persons to be employed?

   (b) How many persons will be employed in the 20th week if the present rate of expansion continues?

11 Car production in the first week of a new model was 150. If it continues with a fall per week of 2%,

   (a) find the total output in 52 weeks from the commencement of production in the new model,

   (b) find how many cars will be produced in the next 52 weeks.

Three-Dimensional Triangulation Problems

OBJECTIVES

1 State the sine and cosine rules for a labelled triangle.
2 Recognise the conditions under which each rule can be used.
3 State and use the formulae for the area of a triangle.
4 Define angles of elevation and depression and apply to problems.
5 Define the angle between a line and a plane.
6 Define the angle between two intersecting planes.
7 Identify relevant planes in a given three-dimensional problem.
8 Solve a three-dimensional triangulation problem capable of being specified within a rectangular prism.

The Solution of Plane Triangles

We have met previously the *sine* and *cosine* rules for finding the sides and angles of non-right-angled triangles, and also three formulae for calculating the areas of triangles.

Fig. 4.1 shows a triangle labelled conventionally, and for convenience the formulae are listed below:

The *sine* rule: $$\frac{a}{\sin A} = \frac{b}{\sin B} = \frac{c}{\sin C}$$

This is used when given: one side and any two angles,

or: two sides and an angle opposite one of the sides.

Fig. **4.1**

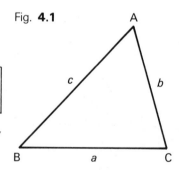

48

The *cosine* rule:

$$a^2 = b^2 + c^2 - 2bc \cos A$$

or:

$$b^2 = a^2 + c^2 - 2ac \cos B$$

or:

$$c^2 = a^2 + b^2 - 2ab \cos C$$

This is used when given: two sides and the angle between them

or: the three sides.

The area of a triangle may be found using:

either | Area $= \frac{1}{2} \times$ base $\times$ altitude

or | Area $= \frac{1}{2}ab \sin C = \frac{1}{2}bc \sin A = \frac{1}{2}ac \sin B$

or | Area $= \sqrt{s(s-a)(s-b)(s-c)}$ where $s = \dfrac{a+b+c}{2}$

The diameter D of the circumscribing circle of a triangle is given by

$$D = \frac{a}{\sin A} = \frac{b}{\sin B} = \frac{c}{\sin C}$$

This topic was covered extensively in *BTEC National NII Mathematics for Technicians* and the worked example which follows should be sufficient revision.

Example 4.1

Find the resultant of the two forces shown in Fig. 4.2 and the angle it makes with the 50 N force.

Fig. **4.2**

Fig. **4.3**

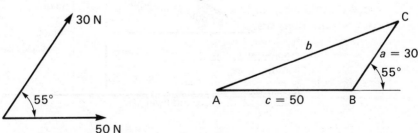

The triangle ABC (Fig. 4.3) is the vector diagram for the given system in which $a = 30$, $c = 50$, and the length b gives the resultant.

Now $\angle ABC = 180° - 55° = 125°$

To find b we will use the cosine rule which gives

$$b^2 = a^2 + c^2 - 2ac \cos B$$

$\therefore \qquad\qquad b^2 = 30^2 + 50^2 - 2 \times 30 \times 50 \times \cos 125°$

$\therefore \qquad\qquad b = 71.56$

To find angle A we will use the sine rule which gives:

$$\frac{a}{\sin A} = \frac{b}{\sin B}$$

from which $\qquad \sin A = \dfrac{a(\sin B)}{b}$

$$= \frac{30(\sin 125°)}{71.56}$$

$$A = 20.08°$$

The resultant is 71.56 N and the angle it makes with the 50 N force is 20.08°.

Angle of Elevation

If you look upwards at an object the angle formed between the horizontal and your line of sight is called the **angle of elevation** (Fig. 4.4).

Fig. **4.4**

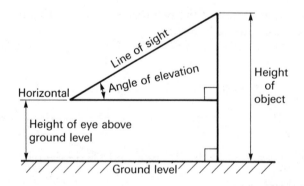

Example 4.2

To find the height of a tower a surveyor sets up his theodolite 100 m from the base of the tower. He finds the angle of elevation of the top of the tower to be 30°. If the instrument is 1.5 m from the ground, what is the height of the tower?

In Fig. 4.5,   Fig. **4.5**

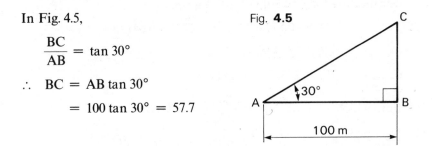

$$\frac{BC}{AB} = \tan 30°$$

$$\therefore \quad BC = AB \tan 30°$$

$$= 100 \tan 30° = 57.7$$

Hence, height of tower $= 57.7 + 1.5 = 59.2$ m.

Example 4.3

To find the height of a pylon, a surveyor sets up a theodolite some distance from the base of the pylon and finds the angle of elevation to the top of the pylon to be 30°. He then moves 60 m nearer to the pylon and finds that the angle of elevation is 42°. Find the height of the pylon assuming that the ground is horizontal and that the theodolite stands 1.5 m above the ground.

Fig. **4.6**

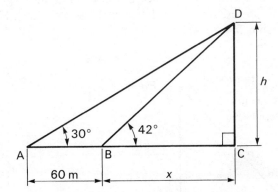

Referring to Fig. 4.6, let $BC = x$ and $DC = h$.

In $\triangle ACD$,   In $\triangle BDC$,

$$\frac{DC}{AC} = \tan 30°$$ $$\frac{DC}{BC} = \tan 42°$$

$$\therefore \quad DC = AC \tan 30°$$ $$\therefore \quad DC = BC \tan 42°$$

or $\quad h = 0.5774(x + 60)$ [1]   or $\quad h = 0.9004x$ [2]

From Equation [2], $x = \dfrac{h}{0.9004} = 1.1106h$

Substituting for x in Equation [1] gives

$$h = 0.5774(1.1106h + 60) = 0.6413h + 34.64$$

$$h - 0.6413h = 34.64$$

or $0.3587h = 34.64$

from which $h = \dfrac{34.64}{0.3587} = 96.6 \text{ m}$

Hence the height of the pylon is $96.6 + 1.5 = 98.1$ m.

Angle of Depression

If you look down at an object, the angle formed between the horizontal and your line of sight is called the *angle of depression* (Fig. 4.7).

Example 4.4

From the top floor window of a house, 14 m above ground level, the angle of depression of an object in the street is 52°. How far is the object from the house?

Fig. **4.7**   Fig. **4.8**

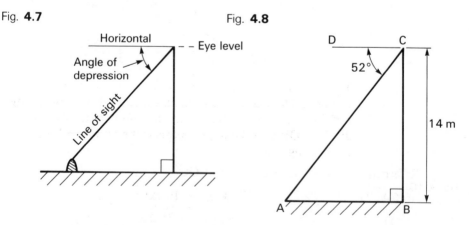

The conditions are shown in Fig. 4.8. Since the angle of depression is 52°, $\angle ACD = 52°$.

$$\angle ACB = 90° - 52° = 38°$$

In $\triangle ABC$, $\dfrac{AB}{CB} = \tan 38°$

$\therefore$ $AB = CB \tan 38° = 14 \tan 38° = 10.9$

Hence the object is 10.9 m from the house.

EXERCISE

4·1

1 The line diagram of a jib crane is shown in Fig. 4.9. Calculate the length of the jib BC and the angle between the jib BC and the tie rod AC.

Fig. **4.9**

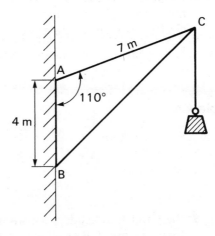

2 The schematic layout of a petrol engine is shown in Fig. 4.10. The crank OC is 80 mm long and the connecting rod CP is 235 mm long. For the position shown find the distance OP and the angle between the crank and the connecting rod.

Fig. **4.10**

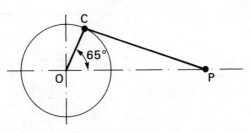

3 A four-bar mechanism ABCD is shown in Fig. 4.11. Links AB, BC and CD are pin-jointed at their ends. (The name 'four-bar' is given because the length AD is considered to be a link.) Find the length AC and the angle ADC.

Fig. **4.11**

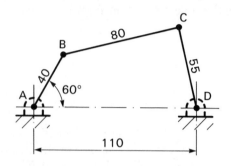

4 Fig. 4.12 shows the dimensions between three holes which are to be drilled in a plate of steel. Find the radius of the circle on which the holes lie.

Fig. **4.12**

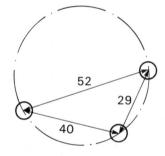

5 P, Q and R are sliders constrained to move along centre-lines OA and OB as shown in Fig. 4.13. For the given position find the distance OR.

Fig. **4.13**

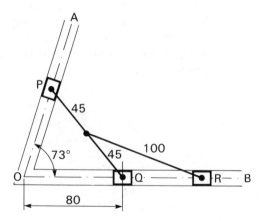

6 Find the lengths of the members BF and CF in the roof truss shown in Fig. 4.14.

Fig. **4.14**

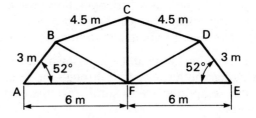

7 Find the area of triangle ABC in Fig. 4.9.

8 Find the area of triangle OCP in Fig. 4.10.

9 Find the area of quadrilateral ABCD in Fig. 4.11.

10 Find the area of the triangle in Fig. 4.12.

11 A man whose eye level is 1.5 m above ground level is 15 m away from a tower 20 m tall. Determine the angle of elevation of the top of the tower from his eyes.

12 A man standing on top of a mountain 1200 m high observes the angle of depression of the top of a steeple to be 43°. If the height of the steeple is 50 m, how far is it from the mountain?

13 To find the height of a tower a surveyor stands some distance away from its base and he observes the angle of elevation to the top of the tower to be 45°. He then moves 80 m nearer to the tower and then finds the angle of elevation to be 60°. If the theodolite, which is used to measure the angles, stands 1.5 m above ground level find the height of the tower.

Three-dimensional Triangulation Problems

Since most components, assemblies, structures, etc. in engineering are three-dimensional it is inevitable that we will have to solve problems of this type. These problems may involve finding angles between lines, true lengths of lines, and even the angle between two planes.

This topic is also useful because it helps us visualise problems in three dimensions, which is an asset for all good engineers.

Most three-dimensional triangulation problems may be solved using only basic trigonometrical ratios and the theorem of Pythagoras for suitably selected right-angled triangles.

No specific method can be laid down for these problems but we feel that, after you have worked through the examples in the text, you will be able to approach with confidence the exercises at the end of the chapter.

Example 4.5

Find the angle BAC in the triangular prism shown in Fig. 4.15.

On inspection it may be seen that several of the faces of the prism are right-angled triangles. Some of these triangles are used in the following solution.

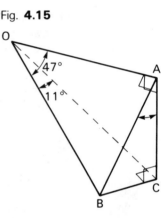

Fig. **4.15**

Using the right-angled triangle OAB

we have, $\sin 47° = \dfrac{AB}{OB}$

∴ $AB = OB \sin 47°$

Using the right-angled triangle OCB

we have, $\sin 11° = \dfrac{BC}{OB}$

∴ $BC = OB \sin 11°$

Using the right-angled triangle ABC

we have $\sin B\widehat{A}C = \dfrac{BC}{AB} = \dfrac{OB \sin 11°}{OB \sin 47°} = \dfrac{\sin 11°}{\sin 47°}$

hence $B\widehat{A}C = 15.12°$

Example 4.6

Fig. 4.16 shows a component which has a rectangular base. The top edge is parallel to the base and one end of the triangular end faces is perpendicular to the base. Find the area of the sloping triangular end face.

Fig. **4.16**

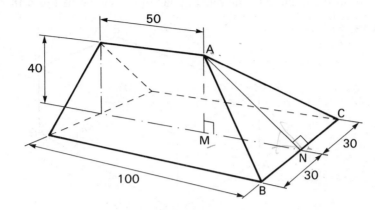

Two construction lines have been drawn. AM is perpendicular to the rectangular base and AN is perpendicular to the edge BC.

From the right-angled triangle AMN and using the theorem of Pythagoras

we have $\quad AN^2 = AM^2 + MN^2$

$$= 40^2 + 50^2 = 4100$$

$\therefore \qquad\qquad AN = 64.03 \text{ mm}$

Now the area of the triangular sloping face $ABC = \frac{1}{2} \times BC \times AN$

$$= \frac{1}{2} \times 60 \times 64.03$$

$$= 1921 \text{ mm}^2$$

The Angle Between a Line and a Plane

In Fig. 4.17 the line AP intersects the xy plane at A. To find the angle between AP and the plane, draw PM perpendicular to the plane and then join AM.

Fig. **4.17**

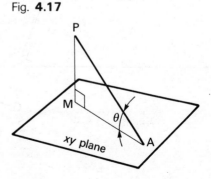

The angle between the line and the plane is $\angle PAM$.

Example 4.7

Find the angle θ in the rectangular block shown in Fig. 4.18.

Fig. **4.18**

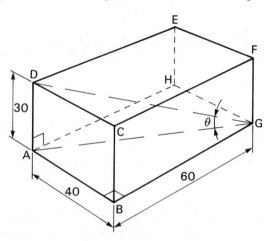

From the right-angled triangle ABG and using the theorem of Pythagoras

we have
$$AG^2 = AB^2 + BG^2$$
$$= 40^2 + 60^2 = 5200$$
$\therefore$
$$AG = 72.11 \text{ mm}$$

In right-angled triangle AGD

we have
$$\tan \theta = \frac{AD}{AG} = \frac{30}{72.11} = 0.4160$$
$\therefore$
$$\theta = 22.59°$$

The Angle Between Two Planes

Fig. **4.19**

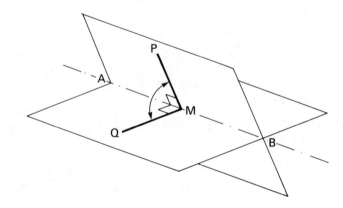

The problem of finding the angle between two planes resolves itself into that of finding an angle between two lines, as shown in Fig. 4.19.

Each of the planes contains one of the lines which meet at the intersection AB of the planes. Also each line is at right angles to AB, that is both PM and QM are perpendicular to the line of intersection AB of the planes. The true angle between the planes is the angle PMQ.

Example 4.8

A right pyramid has a square base 60 mm square, and a height of 50 mm. Find (a) the angle between the base and a sloping face, and (b) the angle between two adjacent sloping faces.

Fig. **4.20**

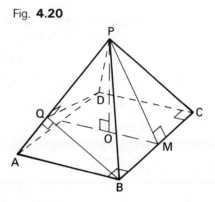

The pyramid is shown suitably labelled in Fig. 4.20.

(a) BC is the line of intersection of a sloping side and the base. PO is the vertical centre line. If OM is drawn perpendicular to BC, then PM is also perpendicular to BC. Hence the angle PMO is the required angle between the base and a sloping side.

Now using the right-angled triangle POM

we have $\tan P\widehat{M}O = \dfrac{PO}{OM} = \dfrac{50}{30} = 1.667$

$\therefore \qquad P\widehat{M}O = 59.04°$

(b) AP is the line of intersection of two adjacent sloping faces. BQ is drawn perpendicular to AP. Then DQ is also perpendicular to AP. Hence angle BQD is the required angle between two adjacent sloping faces.

Now using the right-angled triangle POM and the theorem of Pythagoras we have

$$PM^2 = PO^2 + OM^2 = 50^2 + 30^2 = 3400$$

$\therefore \qquad PM = 58.31 \text{ mm}$

In right-angled triangle PMB

we have $\tan \widehat{PBM} = \dfrac{PM}{BM} = \dfrac{58.31}{30} = 1.944$

∴ $\widehat{PBM} = 62.78°$

From the symmetry of the figure $\widehat{QAB} = \widehat{PBM} = 62.78°$

In right-angled triangle AQB

we have $QB = AB \sin \widehat{QAB}$

$= 60 \sin 62.78°$

$= 53.36 \text{ mm}$

From the right-angled triangle BCD and using the theorem of Pythagoras we have

$$BD^2 = DC^2 + BC^2 = 60^2 + 60^2 = 7200$$

∴ $BD = 84.85 \text{ mm}$

Hence $BO = \tfrac{1}{2}BD = \tfrac{1}{2}(84.85) = 42.43 \text{ mm}$

From the right-angled triangle BOQ

we have $\sin \widehat{BQO} = \dfrac{BO}{QB} = \dfrac{42.43}{53.36} = 0.7952$

∴ $\widehat{BQO} = 52.67°$

But $\widehat{BQD} = 2(\widehat{BQO}) = 2(52.67°)$

∴ $\widehat{BQD} = 105.34°$

Example 4.9

A triangular prism OABC is cut off from a rectangular block by an oblique plane OAC as shown in Fig. 4.21.

 Fig. **4.21**

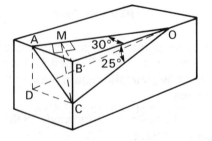

Find (a) the angle between the oblique plane OAC and the top surface of the block, and (b) the angle COD where AD is perpendicular to the top face and CD is perpendicular to the front face of the block.

(a) Draw BM perpendicular to AO and then join CM, which will also be perpendicular to AO. Since both these lines are perpendicular to the line of intersection AO of the plane OAC and the top face of the block, angle CMB is the required angle.

Again, we shall use several right-angled triangles to find this angle.

Using the right-angled triangle BMO,

we have $\qquad \sin 30° = \dfrac{BM}{OB}$

$\therefore \qquad\qquad BM = OB \sin 30°$

Using the right-angled triangle OBC,

we have $\qquad \tan 25° = \dfrac{BC}{OB}$

$\therefore \qquad\qquad BC = OB \tan 25°$

Using the right-angled triangle CMB,

we have $\qquad \tan C\widehat{M}B = \dfrac{BC}{BM} = \dfrac{OB \tan 25°}{OB \sin 30°} = \dfrac{\tan 25°}{\sin 30°}$

$\therefore \qquad\qquad C\widehat{M}B = 43°$

(b) Using the right-angled triangle ABO,

we have $\qquad \tan 30° = \dfrac{AB}{OB}$

$\therefore \qquad\qquad AB = OB \tan 30°$

but ABCD is a rectangle, and hence DC = AB

$\therefore \qquad\qquad DC = AB = OB \tan 30°$

Using the right-angled triangle OBC,

we have $\qquad \cos 25° = \dfrac{OB}{OC}$

$\therefore \qquad\qquad OC = OB \sec 25°$

Using the right-angled triangle OCD,

we have $\qquad \tan C\widehat{O}D = \dfrac{DC}{CO} = \dfrac{OB \tan 30°}{OB \sec 25°} = (\tan 30°)(\cos 25°)$

$\therefore \qquad\qquad C\widehat{O}D = 27.62°$

EXERCISE

4·2

1 Fig. 4.22 shows a small component with a horizontal rectangular base ABCD in which AD is 70 mm and CD is 40 mm. The end face of the component AXB is vertical and it is an equilateral triangle. The top edge XY is horizontal and it is 45 mm long. Find:

(a) The altitude of △YDC (YF in the diagram).
(b) The area of the sloping face YDC.
(c) The angle which the sloping face YDC makes with the base.

Fig. **4.22**

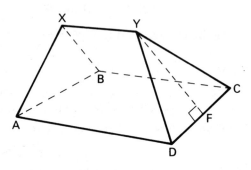

2 Fig. 4.23 shows a triangular prism in which HM is vertical and face ABM is horizontal. The edge HB slopes at 15° to the horizontal. If AB is 55 mm, ∠HBA = 90° and ∠HAB = 55°, find:

(a) The length HB.
(b) The vertical height HM.
(c) The angle HAM which AH makes with the horizontal.

Fig. **4.23**

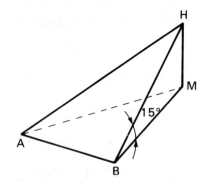

3 The base of the wedge shown in Fig. 4.24 is a rectangle 80 mm long and 60 mm wide. The vertical faces ABC and PQR are equilateral triangles. Calculate:

(a) The angle between the diagonals PB and PC.

(b) The angle between the diagonal PC and the base.

Fig. **4.24**

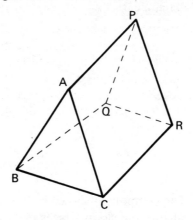

4   Fig. 4.25 shows an elevator hopper with a square back. Find the area of metal required to make it.

Fig. **4.25**

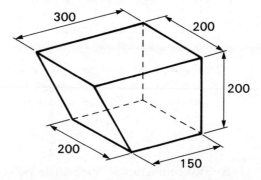

5   Fig. 4.26 shows, in plan view, the roof of a rectangular building, the sloping faces of which are inclined at 34° to the horizontal. Calculate:

(a) The height of the ridge XY above the horizontal plane ABCD.

(b) The angle which the edge AX makes with the horizontal.

(c) The length of AX.

Fig. **4.26**

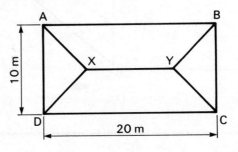

6   Fig. 4.27 shows a rectangular block 70 mm long, 50 mm wide and 40 mm high. Find the lengths of AG and GD and the angle θ.

Fig. **4.27**

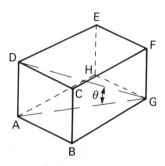

7   The component shown in Fig. 4.28 has a square base and each of the sloping faces makes an angle of 30° with the vertical. Calculate:

(a) The angle between the edge AB and the base of the block.

(b) The true angle between two adjacent sloping faces.

Fig. **4.28**

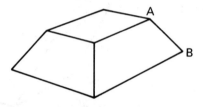

8   Two orthographic third angle views of a symmetrical triangular pyramid are shown in Fig. 4.29. Calculate the true angle BAC.

Fig. **4.29**

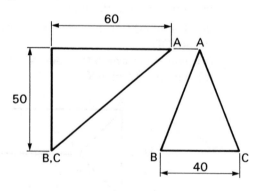

5

The Binomial Theorem

OBJECTIVES

1 Expand expressions of the form $(a + x)^n$ for small, positive integers n.
2 State the general form for the binomial coefficients for all positive integers n.
3 Expand expressions of the form $(1 + x)^n$ where n takes positive, negative or fractional values.
4 State the range of values of x for which the series is convergent.
5 Calculate the effect on the subject of a formula when one or more of the independent variables is subject to a small change or error.

Binomial Expression

A **binomial expression** consists of two terms. Thus $1 + x$, $a + b$, $5y - 2$, $3x^2 + 7$ and $7a^3 + 3b^2$ are all binomial expressions. The **binomial theorem** allows us to expand powers of such expressions.

The Binomial Theorem

Now $(a + b)^0 = 1$

(since any number to the power 0 is unity),

and also $(a + b)^1 = a + b$

Multiplying both sides by $(a + b)$ gives

$$(a + b)^2 = a^2 + 2ab + b^2$$

also $(a + b)^3 = a^3 + 3a^2b + 3ab^2 + b^3$

and $(a + b)^4 = a^4 + 4a^3b + 6a^2b^2 + 4ab^3 + b^4$

We can now arrange the coefficients of each of the terms of the above expansions in the form known as **Pascal's triangle.**

65

Binomial expression	Coefficients in the expansion
$(a+b)^0$	1
$(a+b)^1$	1 1
$(a+b)^2$	1 2 1
$(a+b)^3$	1 3 3 1
$(a+b)^4$	1 4 6 4 1
$(a+b)^5$	1 5 10 10 5 1
$(a+b)^6$	1 6 15 20 15 6 1
$(a+b)^7$	1 7 21 35 35 21 7 1

It will be seen that:

(a) The number of terms in each expansion is one more than the index. Thus the expansion of $(a+b)^9$ will have 10 terms.

(b) The arrangement of the coefficients is symmetrical.

(c) The coefficients of the first and last terms are both always unity.

(d) Each coefficient in the table is obtained by adding together the two coefficients in the line above that lie on either side of it.

The expansion of $(a+b)^8$ is therefore:

$$a^8 + (1+7)a^7b + (7+21)a^6b^2 + (21+35)a^5b^3 + (35+35)a^4b^4$$
$$+ (35+21)a^3b^5 + (21+7)a^2b^6 + (7+1)ab^7 + b^8$$

$$= a^8 + 8a^7b + 28a^6b^2 + 56a^5b^3 + 70a^4b^4 + 56a^3b^5 + 28a^2b^6$$
$$+ 8ab^7 + b^8$$

It is inconvenient to use Pascal's triangle when expanding higher powers of $(a+b)$. In such cases the following series is used:

$$(a+b)^n = a^n + na^{n-1}b + \frac{n(n-1)}{2!}a^{n-2}b^2 + \frac{n(n-1)(n-2)}{3!}a^{n-3}b^3$$
$$+\ldots+b^n$$

This is the **binomial theorem** and is true for all positive whole numbers n.

The symbol '!' indicates 'factorial' when following a positive whole number.

For example, 4! is pronounced 'factorial four' and means $4 \times 3 \times 2 \times 1$.

Thus $2! = 2 \times 1$ $3! = 3 \times 2 \times 1$
        $4! = 4 \times 3 \times 2 \times 1$ $5! = 5 \times 4 \times 3 \times 2 \times 1$
        and so on.

Example 5.1

Expand $(3x + 2y)^4$.

Comparing $(3x + 2y)^4$ with $(a + b)^4$, we have $3x$ in place of a, and $2y$ in place of b. Substituting in the standard expansion, we get

$$[(3x) + (2y)]^4 = (3x)^4 + 4(3x)^3(2y) + \frac{(4)(4 - 1)}{2!}(3x)^2(2y)^2$$

$$+ \frac{(4)(4 - 1)(4 - 2)}{3!}(3x)(2y)^3 + (2y)^4$$

$$= (3x)^4 + 4(3x)^3 2y + \frac{4 \times 3}{2 \times 1}(3x)^2(2y)^2$$

$$+ \frac{4 \times 3 \times 2}{3 \times 2 \times 1}(3x)(2y)^3 + (2y)^4$$

$$= 81x^4 + 216x^3y + 216x^2y^2 + 96xy^3 + 16y^4$$

Example 5.2

Expand $(x - 4y)^{15}$ to four terms.

The given binomial expression should be rewritten as $[x + (-4y)]^{15}$, and comparing this with the standard expression for $(a + b)^n$ we have x in place of a, $-4y$ in place of b, and 15 in place of n.

$$\therefore [x + (-4y)]^{15} = x^{15} + 15x^{14}(-4y) + \frac{(15)(15 - 1)}{2!}x^{13}(-4y)^2$$

$$+ \frac{15(15 - 1)(15 - 2)}{3!}x^{12}(-4y)^3 + \ldots$$

$$= x^{15} + 15(-4)x^{14}y + \frac{15 \times 14 \times (-4)^2}{2 \times 1}x^{13}y^2$$

$$+ \frac{15 \times 14 \times 13 \times (-4)^3}{3 \times 2 \times 1}x^{12}y^3 + \ldots$$

$$= x^{15} - 60x^{14}y + 1680x^{13}y^2 - 29\,120x^{12}y^3 + \ldots$$

EXERCISE

5·1

    1 $(1+z)^5$ **2** $(p+q)^6$ **3** $(x-3y)^4$

    4 $(2p-q)^5$ **5** $(2x+y)^7$ **6** $\left(x+\dfrac{1}{x}\right)^3$

Expand to four terms using the binomial theorem:

    7 $(1+x)^{12}$ **8** $(1-2x)^{14}$ **9** $(p+q)^{16}$

  10 $(1+3y)^{10}$ **11** $(x^2-3y)^9$ **12** $\left(x^2+\dfrac{1}{x^2}\right)^{11}$

The Binomial Series

If we put $a = 1$ and use x instead of b in the binomial theorem we get

$$(1+x)^n = 1+nx+\frac{n(n-1)}{2!}x^2+\frac{n(n-1)(n-2)}{3!}x^3+\ldots+x^n$$

This expression is true for all positive whole number values of n.

For negative and fractional values of n the right-hand side of the above expression has no final term. It has an infinite number of terms and is called an infinite series and forms the binomial series:

$$(1+x)^n = 1+nx+\frac{n(n-1)}{2!}x^2+\frac{n(n-1)(n-2)}{3!}x^3+\ldots$$

There is also a proviso that, for negative and fractional values of n, any value given to x must lie between $+1$ and -1. The reason for this is explained under the heading '**Convergence of Series**' on p. 70.

Example 5.3

Use the binomial series to expand $\sqrt{(1+x)}$ to five terms.

$$\sqrt{(1+x)} = (1+x)^{1/2}$$

$$= 1+\tfrac{1}{2}x+\frac{\tfrac{1}{2}(\tfrac{1}{2}-1)}{2!}x^2+\frac{\tfrac{1}{2}(\tfrac{1}{2}-1)(\tfrac{1}{2}-2)}{3!}x^3$$

$$+\frac{\tfrac{1}{2}(\tfrac{1}{2}-1)(\tfrac{1}{2}-2)(\tfrac{1}{2}-3)}{4!}x^4+\ldots$$

$$= 1 + \tfrac{1}{2}x + \frac{(\tfrac{1}{2})(-\tfrac{1}{2})}{2 \times 1}x^2 + \frac{\tfrac{1}{2}(-\tfrac{1}{2})(-\tfrac{3}{2})}{3 \times 2 \times 1}x^3$$

$$+ \frac{\tfrac{1}{2}(-\tfrac{1}{2})(-\tfrac{3}{2})(-\tfrac{5}{2})}{4 \times 3 \times 2 \times 1}x^4 + \ldots$$

$$= 1 + \tfrac{1}{2}x - \frac{1}{2 \times 2 \times 2}x^2 + \frac{3}{2 \times 2 \times 2 \times 3 \times 2 \times 1}x^3$$

$$- \frac{3 \times 5}{2 \times 2 \times 2 \times 2 \times 4 \times 3 \times 2 \times 1}x^4 + \ldots$$

$$= 1 + \tfrac{1}{2}x - \tfrac{1}{8}x^2 + \tfrac{1}{16}x^3 - \tfrac{5}{128}x^4 + \ldots$$

Example 5.4

Use the binomial series to expand $\dfrac{1}{(1 + 2x)^5}$ to four terms.

$$\frac{1}{(1 + 2x)^5} = [1 + (2x)]^{-5}$$

$$= 1 + (-5)(2x) + \frac{(-5)(-5-1)}{2!}(2x)^2$$

$$+ \frac{(-5)(-5-1)(-5-2)}{3!}(2x)^3 + \ldots$$

$$= 1 - 10x + \frac{5 \times 6}{2 \times 1} \times 4x^2 - \frac{5 \times 6 \times 7}{3 \times 2 \times 1} \times 8x^3 + \ldots$$

$$= 1 - 10x + 60x^2 - 280x^3 + \ldots$$

Example 5.5

Use the binomial series to expand $\dfrac{1}{\sqrt[4]{(1 - x)}}$ to four terms.

$$\frac{1}{\sqrt[4]{(1 - x)}} = [1 - x]^{-1/4}$$

$$= [1 + (-x)]^{-1/4}$$

$$= 1 + (-\tfrac{1}{4})(-x) + \frac{(-\tfrac{1}{4})(-\tfrac{1}{4} - 1)}{2!}(-x)^2$$

$$+ \frac{(-\tfrac{1}{4})(-\tfrac{1}{4} - 1)(-\tfrac{1}{4} - 2)}{3!}(-x)^3 + \dots$$

$$= 1 + \tfrac{1}{4}x + \frac{(-\tfrac{1}{4})(-\tfrac{5}{4})}{2 \times 1}x^2 - \frac{(-\tfrac{1}{4})(-\tfrac{5}{4})(-\tfrac{9}{4})}{3 \times 2 \times 1}x^3 + \dots$$

$$= 1 + \tfrac{1}{4}x + \frac{5}{4 \times 4 \times 2}x^2 + \frac{5 \times 9}{4 \times 4 \times 4 \times 3 \times 2}x^3 + \dots$$

$$= 1 + \tfrac{1}{4}x + \tfrac{5}{32}x^2 + \tfrac{15}{128}x^3 + \dots$$

Convergence of Series

Consider the binomial series obtained in Example 5.3 which gave

$$(1 + x)^{1/2} = 1 + \tfrac{1}{2}x - \tfrac{1}{8}x^2 + \tfrac{1}{16}x^3 - \tfrac{5}{128}x^4 + \dots$$

Putting $x = \tfrac{1}{2}$ we get

$$(1 + \tfrac{1}{2})^{1/2} = 1 + \tfrac{1}{2}(\tfrac{1}{2}) - \tfrac{1}{8}(\tfrac{1}{2})^2 + \tfrac{1}{16}(\tfrac{1}{2})^3 - \tfrac{5}{128}(\tfrac{1}{2})^4 + \dots$$

$$= 1 + 0.25 - 0.031\,25 + 0.007\,813 - 0.002\,441 + \dots$$

$$= 1.22 \text{ correct to 3 significant figures}$$

We can see that the terms in the expansion when $x = \tfrac{1}{2}$ are rapidly getting smaller and a numerical answer to any reasonable degree of accuracy may soon be obtained.

Such a series is said to be **convergent**.

Let us now put $x = 2$ into the same expansion, getting

$$(1 + 2)^{1/2} = 1 + \tfrac{1}{2}(2) - \tfrac{1}{8}(2)^2 + \tfrac{1}{16}(2)^3 - \tfrac{5}{128}(2)^4 + \dots$$

$$= 1 + 1 - 0.5 + 0.5 - 0.625 + \dots$$

In this series the terms do not get progressively smaller — in fact they subsequently become larger. This means it is not possible to use the series to find a numerical value and the series is *not* convergent.

In general when using the *binomial series*:

> If index n is negative or fractional, then the value of x must lie between $+1$ and -1.

However, if index n is a positive whole number the series is finite and a numerical answer may always be obtained.

EXERCISE

5·2

1 $(1+x)^6$ **2** $(1+2x)^9$ **3** $(1+x)^{-4}$

4 $(1-x)^{-1/2}$ **5** $(1+x)^{-1}$ **6** $(1+2x)^{-3}$

7 $(1-x)^{-3}$ **8** $\sqrt[3]{(1+x^2)}$ **9** $\dfrac{1}{\sqrt{(1-3x)}}$

10 $\left(1-\dfrac{2x}{3}\right)^{3/4}$

Application to Small Errors

The binomial series is

$$(1+x)^n = 1 + nx + \frac{n(n-1)}{2!}x^2 + \frac{n(n-1)(n-2)}{3!}x^3 + \ldots$$

When x is small compared with unity, e.g. 0.03, then

$$(1+x)^n = 1 + nx \quad \text{approximately}$$

as all other terms in the series contain powers of x which are negligible when compared with the first two shown. This relationship is often useful.

It is easy enough to measure the expansion of a metal rod which has been heated. It is, however, not so easy to measure the increases in areas and volumes which have been expanded.

If α is the coefficient of linear expansion, then the length of an expanded bar originally 1 m long is $(1+\alpha)$ m for one degree temperature increase. An area of the same material originally 1 m × 1 m becomes $(1+\alpha)^2$ m² when it expands. From the above approximation, as α is very small,

$$(1+\alpha)^2 \approx 1 + 2\alpha$$

Hence *the area coefficient is approximately twice the linear coefficient.*

Similarly a volume 1 m × 1 m × 1 m on expanding becomes $(1+\alpha)^3$ m³ and again, using the above approximation,

$$(1+\alpha)^3 \approx 1 + 3\alpha$$

which shows that *the volume coefficient is approximately three times the linear coefficient.*

Example 5.6

In measuring the radius of a circle, the measurement is 1% too large. If this measurement is used to calculate the area of the circle find the resulting error.

Let A be the area of the circle, and r the radius. If δA is the error in the area, then

$$A = \pi r^2$$

and

$$A + \delta A = \pi(r + r \times \tfrac{1}{100})^2$$

$$= \pi r^2(1 + \tfrac{1}{100})^2$$

and since $\tfrac{1}{100}$ is small compared with 1, then

$$A + \delta A \approx \pi r^2(1 + 2 \times \tfrac{1}{100})$$

$$\approx A(1 + \tfrac{2}{100})$$

$$\therefore \qquad \delta A \approx \tfrac{2}{100}A$$

Hence

$$\delta A \approx 2\% \text{ of } A$$

Thus an error of 1% too large in the radius gives an approximate error of 2% too large in the area.

Example 5.7

When a uniform beam is simply supported at its ends the deflection at the centre of span is given by

$$y = \frac{5wl^4}{384EI}$$

where w is the distributed load per unit length, l is the length between supports, E is Young's modulus and I is the 2nd moment of cross-sectional area. Find the percentage change in y when l decreases by 3%.

Let the change in y be δy. Then

$$y + \delta y = \frac{5w}{384EI}[l - l \times \tfrac{3}{100}]^4$$

$$= \frac{5wl^4}{384EI} [1 + (-\tfrac{3}{100})]^4$$

$$= \frac{5wl^4}{384EI} [1 + 4(-\tfrac{3}{100})] \quad \text{approximately}$$

since all higher powers of $\frac{3}{100}$ are negligible.

Thus $\qquad y + \delta y \approx y[1 - \tfrac{12}{100}]$

$\therefore \qquad\qquad \delta y \approx -\tfrac{12}{100}y \quad \text{or} \quad 12\% \text{ of } y \text{ (decrease)}$

Hence y will decrease by 12% approximately.

Example 5.8

A formula used in connection with close-coiled helical springs is

$$P = \frac{GFd^5}{8hD^3}$$

Find the change in P if d is increased by 2% and D is decreased by 3%.

Let δP be the change in P. Then

$$P + \delta P = \frac{GF[d + \tfrac{2}{100}d]^5}{8h[D - \tfrac{3}{100}D]^3}$$

$$= \frac{GFd^5[1 + \tfrac{2}{100}]^5}{8hD^3[1 - \tfrac{3}{100}]^3}$$

$$= P[1 + (\tfrac{2}{100})]^5[1 + (-\tfrac{3}{100})]^{-3}$$

$$\approx P[1 + 5(\tfrac{2}{100})][1 + (-3)(-\tfrac{3}{100})]$$

since all higher powers of $\frac{2}{100}$ and $\frac{3}{100}$ are negligible.

Thus $\qquad P + \delta P \approx P[1 + \tfrac{10}{100}][1 + \tfrac{9}{100}]$

$$= P[1 + \tfrac{19}{100}] \quad \text{approximately}$$

$$= P + \tfrac{19}{100}P$$

$\therefore \qquad\qquad \delta P \approx \tfrac{19}{100}P = 19\% \text{ of } P$

Hence P will increase by 19% approximately.

EXERCISE

5·3

1 Show that an error of 1% in the measurement of the radius of a sphere leads to an error of approximately 2% in the outer surface area, and 3% in the volume.

2 Find the approximate percentage error in the calculated volume of a right circular cone if the radius is taken as 2% too large, and the height is taken as 3% too small.

3 A formula used in connection with helical springs is $y = \dfrac{8WnD^3}{Gd^4}$. Find the percentage error in y if D is 1% too small, and d is 2% too large.

4 In the standard gas equation $\dfrac{pV}{T} = k$, the volume V is increased by 2%, the temperature T is diminished by 1%, and the value of constant k remains unaltered. What is the corresponding percentage change in the pressure p?

5 The deflection y at the centre of a steel rod of length l and circular cross-section of diameter d, simply supported at its ends and carrying a concentrated load W at its centre, is given by the formula $y = \dfrac{Wl^3}{d^4}$. Find the percentage change in y when l increases by 2% and d decreases by 3%.

6 If x is so small that x^3 and all other higher powers of x can be neglected, find an approximation of $3/(\sqrt{2+x})$.

7 If the height h of an isosceles triangle is small compared with the base of length $2b$, show that each of the slant sides has a length of approximately $b + h^2/2b$.

8 The volume per hour of water V passing through an injector of diameter d due to a pressure p is given by the formula $V = 2d^2p^{1/8}$. Find the approximate change in d if p is increased by 5% and V is decreased by 3.5%.

9 The resonant frequency of an oscillation in electrical circuits is given by the formula $f = 1/(2\pi\sqrt{LC})$. If the error in measuring L is 2%, and that in measuring C is 1%, calculate the maximum percentage error in calculating f.

10 A taut wire is held horizontally at two fixed points $2L$ inches apart. If the centre of the wire is deflected vertically by a small distance x, use the binomial theorem to show that the wire extends by an amount approximately equal to $x^2/2L$.

11 The period t of oscillation of a simple pendulum is given by the formula $t = 2\pi\sqrt{l/g}$ where l is the length and g the acceleration due to gravity. If, in an experiment, l is measured 1% too large and g is 5% too large, find the approximate error in t.

The Exponential Function

OBJECTIVES

1 State the expansion of e^x as a power series.

2 Deduce the expansion of e^{-x}

3 Appreciate that the expansions are convergent for all values of x.

4 Deduce an expansion for ae^{kx} where k is positive or negative.

5 Deduce the series for e and evaluate e correct to four decimal places.

6 Use the series of e^x to deduce the unique property $\dfrac{d}{dx}(e^x) = e^x$.

Definition

The term **the exponential function** is generally used to mean the function e^x, where e is the base of Napierian logarithms.

The Exponential Series

The function e^x may be expanded as a power series and this is called the **exponential series**.

The exponential series is

$$e^x = 1 + x + \frac{x^2}{2!} + \frac{x^3}{3!} + \frac{x^4}{4!} + \frac{x^5}{5!} + \frac{x^6}{6!} + \ldots$$

Convergence

Suppose we wish to evaluate e^2. We put $x = 2$ into the series for e^x.
Thus

$$e^2 = 1 + 2 + \frac{2^2}{2!} + \frac{2^3}{3!} + \frac{2^4}{4!} + \frac{2^5}{5!} + \frac{2^6}{6!} + \ldots$$

$$= 1 + 2 + \frac{2 \times 2}{2 \times 1} + \frac{2 \times 2 \times 2}{3 \times 2 \times 1} + \frac{2 \times 2 \times 2 \times 2}{4 \times 3 \times 2 \times 1} + \frac{2 \times 2 \times 2 \times 2 \times 2}{5 \times 4 \times 3 \times 2 \times 1}$$

$$+ \frac{2 \times 2 \times 2 \times 2 \times 2 \times 2}{6 \times 5 \times 4 \times 3 \times 2 \times 1} + \ldots$$

$$= 1 + 2 + 2 + 1.3333 + 0.6667 + 0.2667 + 0.0889 + \ldots$$

Although initially the terms get larger, the fourth and subsequent terms become successively smaller. Thus the series is *convergent* and may be used to evaluate e^2 to any desired accuracy.

In general the exponential series is convergent for all values of x.

This means the series may be used to find the values of expressions such as $e^{3.1}$, e^{-4}, $e^{-1.56}$, etc. You may verify this by working through Question 1 in Exercise 6.1 at the end of this chapter.

Example 6.1

Find a series for e^{-x}.

If we substitute $-x$ for x in the series for e^x, then

$$e^{-x} = 1 + (-x) + \frac{(-x)^2}{2!} + \frac{(-x)^3}{3!} + \frac{(-x)^4}{4!} + \frac{(-x)^5}{5!} + \frac{(-x)^6}{6!} + \ldots$$

or

$$e^{-x} = 1 - x + \frac{x^2}{2!} - \frac{x^3}{3!} + \frac{x^4}{4!} - \frac{x^5}{5!} + \frac{x^6}{6!} - \ldots$$

Example 6.2

Find a series for $2e^3$.

If we substitute $x = 3$ in the series for e^x, then

$$e^3 = 1 + 3 + \frac{3^2}{2!} + \frac{3^3}{3!} + \frac{3^4}{4!} + \frac{3^5}{5!} + \ldots$$

$$\therefore \qquad 2e^3 = 2\left(1 + 3 + \frac{3^2}{2!} + \frac{3^3}{3!} + \frac{3^4}{4!} + \frac{3^5}{5!} + \ldots\right)$$

The answer is better left in this form unless a numerical value is required.

Example 6.3

Find a series for $\frac{1}{2}(e^{4x} + e^{-4x})$.

If we substitute $4x$ for x in the series for e^x, then

$$e^{4x} = 1 + 4x + \frac{(4x)^2}{2!} + \frac{(4x)^3}{3!} + \frac{(4x)^4}{4!} + \frac{(4x)^5}{5!} + \ldots$$

and if we also substitute $-4x$ for x in the series for e^x, then:

$$e^{-4x} = 1 + (-4x) + \frac{(-4x)^2}{2!} + \frac{(-4x)^3}{3!} + \frac{(-4x)^4}{4!} + \frac{(-4x)^5}{5!} + \ldots$$

$$= 1 - 4x + \frac{(4x)^2}{2!} - \frac{(4x)^3}{3!} + \frac{(4x)^4}{4!} - \frac{(4x)^5}{5!} + \ldots$$

Now adding the series we have

$$e^{4x} + e^{-4x} = 1 + 4x + \frac{(4x)^2}{2!} + \frac{(4x)^3}{3!} + \frac{(4x)^4}{4!} + \frac{(4x)^5}{5!} + \ldots$$

$$+ 1 - 4x + \frac{(4x)^2}{2!} - \frac{(4x)^3}{3!} + \frac{(4x)^4}{4!} - \frac{(4x)^5}{5!} + \ldots$$

$$= 2(1) + 2\left[\frac{(4x)^2}{2!}\right] + 2\left[\frac{(4x)^4}{4!}\right] + 2\left[\frac{(4x)^6}{6!}\right] + \ldots$$

And dividing both sides by 2, we have

$$\tfrac{1}{2}(e^{4x} + e^{-4x}) = 1 + \frac{(4x)^2}{2!} + \frac{(4x)^4}{4!} + \frac{(4x)^6}{6!} + \ldots$$

The Numerical Value of e

Although we have defined the exponential function as e^x, we still do not know how to arrive at the numerical value of the constant e.

This value may be found by putting $x = 1$ in the series for e^x.

We must also decide to what accuracy we require the numerical answer. This will determine how many terms we will have to include in our calculations. Suppose we decide on an answer correct to four decimal places. In the final rounding off we must make sure that the fifth decimal place is correct and this in turn will be affected by the sixth decimal place figures.

Thus

$$e = e^1 = 1 + 1 + \frac{1^2}{2!} + \frac{1^3}{3!} + \frac{1^4}{4!} + \frac{1^5}{5!} + \frac{1^6}{6!} + \frac{1^7}{7!} + \frac{1^8}{8!} + \frac{1^9}{9!} + \ldots$$

$$
\begin{aligned}
= \quad &1.000\,000 \\
+ &1.000\,000 \\
+ &0.500\,000 \quad (\tfrac{1}{2} \text{ of the previous term}) \\
+ &0.166\,667 \quad (\tfrac{1}{3} \text{ of the previous term}) \\
+ &0.041\,667 \quad (\tfrac{1}{4} \text{ of the previous term}) \\
+ &0.008\,333 \quad (\tfrac{1}{5} \text{ of the previous term}) \\
+ &0.001\,389 \quad (\tfrac{1}{6} \text{ of the previous term}) \\
+ &0.000\,198 \quad (\tfrac{1}{7} \text{ of the previous term}) \\
+ &0.000\,025 \quad (\tfrac{1}{8} \text{ of the previous term}) \\
+ &0.000\,003 \quad (\tfrac{1}{9} \text{ of the previous term})
\end{aligned}
$$

$$\therefore \qquad e = \quad 2.718\,282$$

Thus the value of e is 2.7183 correct to 4 decimal places.

The Unique Property $\dfrac{d}{dx}(e^x) = e^x$

We have

$$e^x = 1 + x + \frac{x^2}{2!} + \frac{x^3}{3!} + \frac{x^4}{4!} + \ldots$$

Now, if we differentiate this series term by term we obtain

$$\frac{d}{dx}(e^x) = 1 + \frac{2x}{2!} + \frac{3x^2}{3!} + \frac{4x^3}{4!} + \ldots$$

and considering a typical coefficient, for example,

$$\frac{4}{4!} = \frac{4}{4 \times 3 \times 2 \times 1} = \frac{1}{3!}$$

Thus
$$\frac{d}{dx}(e^x) = 1 + x + \frac{x^2}{2!} + \frac{x^3}{3!} + \ldots$$

which is the same as the original series.

Hence
$$\frac{d}{dx}(e^x) = e^x$$

This is the only mathematical function which does *not* change on differentiation.

To put it another way we may say that for the graph of $y = e^x$ the rate of change $\dfrac{dy}{dx}$, at any point, is equal to e^x,   i.e. $\dfrac{dy}{dx} = y$.

EXERCISE

The following examples should be solved using the exponential series. Numerical answers may be checked using a calculator.

1 Find, correct to three significant figures, the values of

(a) $e^{1.5}$ (b) $e^{3.1}$ (c) e^{-2} (d) $e^{-4.3}$

(e) $e^{0.2}$ (f) $e^{0.53}$ (g) $e^{-0.7}$

2 Find the first four terms of the series for

(a) e^{3x} (b) $e^{0.5x}$ (c) $e^{-1.3x}$ (d) $e^{-0.3x}$

3 Evaluate to three decimal places

(a) $3\left(e - \dfrac{1}{e}\right)$ (b) $\tfrac{1}{2}(e^{0.2} + e^{-0.2})$

4 Find the first four terms of the series for $\dfrac{e^{2x} + 1}{2e^x}$.

Trigonometrical Waveforms

OBJECTIVES

1 State the approximations for sin x, cos x and tan x when x is small.

2 Sketch the graphs of

$$\sin A, \sin 2A, 2\sin A \text{ and } \sin\frac{A}{2} \quad \text{and} \quad \cos A, \cos 2A, 2\cos A \text{ and } \cos\frac{A}{2}$$

for values of A between 0° and 360°.

3 Sketch the graphs of $\sin^2 A$ and $\cos^2 A$ for values of A between 0° and 360°.

4 Distinguish between angular and time bases.

5 Define and identify amplitude and frequency.

6 Define angular velocity and period.

7 Sketch the graphs of the functions in **2** where A is replaced by ωt.

8 Determine the single wave resulting from a combination of two waves of the same frequency using phasors and a graphical method.

9 Define the term 'phase angle'.

10 Measure the amplitude and phase angle of the resultant wave in **8.**

11 Determine graphically the single wave resulting from the combination of two waves (within the limitations of **2**).

12 Show that the resultant of two sine waves of different frequencies gives rise to a non-sinusoidal periodic function.

Radians and Degrees

We know that one full revolution is equivalent to 360° or 2π radians.

Hence

$$1 \text{ radian} = \left(\frac{360}{2\pi}\right)^{\circ} = \left(\frac{180}{\pi}\right)^{\circ}$$

If tables are used when finding trigonometrical ratios, such as sines and cosines of angles, it may be necessary to convert an angle in radians to an angle in degrees.

For example

$$\sin 0.5 = \sin\left(0.5 \times \frac{180}{\pi}\right)^{\circ} = \sin 28.65^{\circ} = 0.4795$$

It should be noted that if the units of an angle are omitted it is assumed that it is given in radians (as in the above example).

If a scientific calculator is available it is often possible to set it to accept radians by using a special key. There is then no necessity to convert from radians to degrees.

Approximations for Trigonometrical Functions of Small Angles

If the angle θ is small, i.e. less than about 0.1 radians or 6°, the following approximations are often used:

$\sin \theta \approx \theta$	
$\cos \theta \approx 1 - \dfrac{\theta^2}{2}$	Providing θ is in *radians*
$\tan \theta \approx \theta$	

For example, when $\theta = 0.1$ rad (5.73°):

Function	Correct to four significant figures	
	True value from calculator	Approximate value
$\sin 0.1$	0.0998	0.1000
$\cos 0.1$	0.9950	$1 - \dfrac{(0.1)^2}{2} = 0.9950$
$\tan 0.1$	0.1003	0.1000

Amplitude or Peak Value

The graphs of $\sin \theta$ and $\cos \theta$ each have a maximum value of $+1$ and a minimum value of -1.

Fig. **7.1**

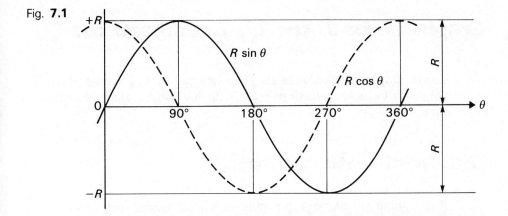

Similarly the graphs of $R \sin \theta$ and $R \cos \theta$ each have a maximum value of $+R$ and a minimum value of $-R$. These graphs are shown in Fig. 7.1.

The value of R is known as the **amplitude** or **peak value**.

Graphs of sin θ, sin 2θ, 2 sin θ and sin $\dfrac{\theta}{2}$

Curves of the above trigonometrical functions are shown plotted in Fig. 7.2. You may find it useful to construct the curves using values obtained from a calculator.

Fig. **7.2**

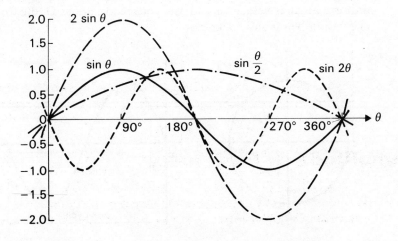

Graphs of cos θ, cos 2θ, 2 cos θ and cos $\dfrac{\theta}{2}$

Cosine graphs are similar in shape to sine curves. You should plot graphs of the above functions from 0° to 360° using values obtained from a calculator.

Graphs of sin² θ and cos² θ

It is sometimes necessary in engineering applications, such as when finding the root mean square value of alternating currents and voltages, to be familiar with the curves $\sin^2 \theta$ and $\cos^2 \theta$.

Values of the functions can be obtained using a calculator and their graphs are shown in Fig. 7.3. We should note that the curves are wholly positive, since squares of negative or positive values are always positive.

Fig. **7.3**

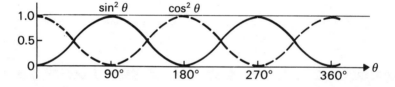

Relation between Angular and Time Scales

In Fig. 7.4 OP represents a radius, of length R, which rotates at a uniform angular velocity ω radians per second about O, the direction of rotation being anticlockwise.

Fig. **7.4**

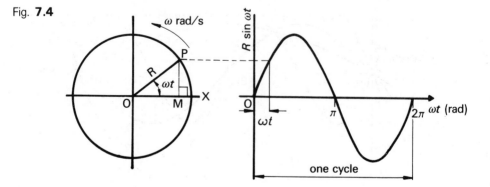

Now

$$\text{Angular velocity} = \frac{\text{Angle turned through}}{\text{Time taken}}$$

∴   Angle turned through = (Angular velocity) × (Time taken)

and hence after a time t seconds

Angle turned through = ωt radians

Also from the right-angled triangle OPM:

$$\frac{PM}{OP} = \sin \widehat{POM}$$

∴ $PM = OP \sin \widehat{POM}$

or $PM = R \sin \omega t$

If a graph is drawn, as in Fig. 7.4, showing how PM varies with the angle ωt, the sine wave representing $R \sin \omega t$ is obtained. It can be seen that the peak value of this sine wave is R (i.e. the magnitude of the rotating radius).

The horizontal scale shows the angle turned through, ωt, and the waveform is said to be plotted on an **angular** or ωt **base**.

Cycle

A cycle is the portion of the waveform which shows its complete shape without any repetition. It may be seen from Fig. 7.4 that one cycle is completed whilst the radius OP turns through 360° or 2π radians.

Period

This is the time taken for the waveform to complete one cycle.

It will also be the time taken for OP to complete one revolution or 2π radians.

Now we know that

$$\text{Time taken} = \frac{\text{Angle turned through}}{\text{Angular velocity}}$$

Hence $\boxed{\text{The period} = \frac{2\pi}{\omega} \text{ seconds}}$

Frequency

The number of cycles per second is called the frequency. The unit of frequency representing one cycle per second is the **hertz** (Hz).

Now if   1   cycle is completed in $\dfrac{2\pi}{\omega}$ seconds (a period)

then $1 \div \dfrac{2\pi}{\omega}$ cycles are completed in 1 second

and therefore $\dfrac{\omega}{2\pi}$ cycles are completed in 1 second

Hence

$$\boxed{\text{Frequency} = \frac{\omega}{2\pi} \text{ Hz}}$$

Also since

$$\text{Period} = \frac{2\pi}{\omega} \text{ s}$$

$$\boxed{\text{Frequency} = \frac{1}{\text{Period}}}$$

Graphs of sin t, sin $2t$, 2 sin t, and sin$\frac{1}{2}t$

Now the waveform sin ωt has a period of $\dfrac{2\pi}{\omega}$ seconds.

Thus the waveform sin t has a period of $\dfrac{2\pi}{1} = 6.28$ seconds.

We have seen how a graph may be plotted on an 'angular' or 'ωt' base as in Fig. 7.4. Alternatively the units on the horizontal axis may be those of time (usually seconds), and this is called a 'time' base, as displayed on an oscilloscope.

In order to plot one complete cycle of the waveform it is necessary to take values of t from 0 to 6.28 seconds. We suggest that you plot the curve of sin t, remembering to set your calculator to the 'radian' mode when finding the value of sin t. The curve is shown plotted on a time base in Fig. 7.5

Similarly,

the waveform sin $2t$ has a period of $\dfrac{2\pi}{2} = 3.14$ seconds

and the waveform sin $\frac{1}{2}t$ has a period of $\dfrac{2\pi}{\frac{1}{2}} = 12.56$ seconds.

Each of the above waveforms has an amplitude of unity. However, the waveform 2 sin t has an amplitude of 2, although its period is the same as that of sin t, namely 6.28 seconds.

All these curves are shown plotted in Fig. 7.5. This enables a visual comparison to be made and it may be seen, for example, that the curve of sin $2t$ has a frequency twice that of sin t (since two cycles of sin $2t$ are completed during one cycle of sin t).

Fig. **7.5**

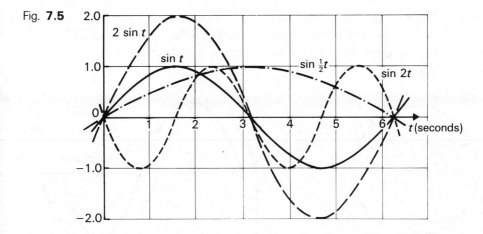

Graphs of *R* cos *ωt*

The waveforms represented by $R \cos \omega t$ are similar to sine waveforms, R being the peak value and $\dfrac{2\pi}{\omega}$ the period. You are left to plot these as instructed in Exercise 7.1 which follows this text.

EXERCISE

7·1

1 On the same axes, using a time base, plot the waveforms of cos t and 2 cos t for one complete cycle from $t = 0$ to $t = 6.28$ seconds.

2 Using the same axes on which the curves were plotted in Question 1, plot the waveforms of cos 2*t* and cos $\dfrac{t}{2}$.

3 On the same axes, using an angle base from 0° to 360°, sketch the following waveforms:

(a) 5 cos θ   (b) 3 sin 2θ   (c) 4 cos 3θ   (d) 2 sin 3θ

Phase Angle

The principal use of sine and cosine waveforms occurs in electrical technology in which they represent alternating currents and voltages. In a diagram such as that shown in Fig. 7.6 the rotating radii OP and OQ are called **phasors**.

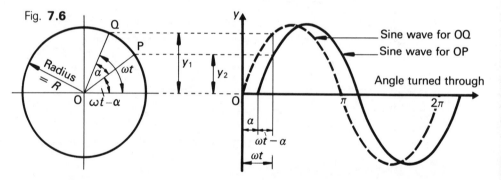

Fig. **7.6**

Fig. 7.6 shows two phasors OP and OQ, separated by an angle α, rotating at the same angular speed in an anticlockwise direction. The sine waves produced by OP and OQ are identical curves but they are displaced from each other. The amount of displacement is known as the **phase difference** and, measured along the horizontal axis, is α. The angle α is called the **phase angle**.

In Fig. 7.6 the phasor OP is said to **lag** behind phasor OQ by the angle α. If the radius of the phasor circle is *R* then OP = OQ = *R* and hence

for the phasor OQ, $y_1 = R \sin \omega t$

and for the phasor OP,

$$y_2 = R \sin (\omega t - \alpha)$$

Fig. **7.7**

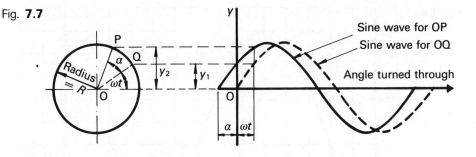

Similarly in Fig. 7.7 the phasor OP **leads** the phasor OQ by the phase angle α.

Hence for the phasor OQ,

$$y_1 = R \sin \omega t$$

and for the phasor OP,

$$y_2 = R \sin (\omega t + \alpha)$$

In practice it is usual to draw waveform on an 'angular' or 'ωt' base when considering phase angles, as in the following example.

Example 7.1

Sketch the waveforms of $\sin \omega t$ and $\sin \left(\omega t - \dfrac{\pi}{3} \right)$ on an angular base and identify the phase angle.

The curve $\sin \omega t$ will be plotted between $\omega t = 0$ and $\omega t = 2\pi$ radians (i.e. over 1 cycle).

Also $\sin \left(\omega t - \dfrac{\pi}{3} \right)$ will be plotted between values given by

$$\omega t - \frac{\pi}{3} = 0 \qquad \text{and when} \quad \omega t - \frac{\pi}{3} = 2\pi$$

i.e. $\quad \omega t = \dfrac{\pi}{3}$ radians $\qquad$ and $\omega t = 2\pi + \dfrac{\pi}{3} = \dfrac{7\pi}{3}$ radians

The graphs are shown in Fig. 7.8.

Fig. **7.8**

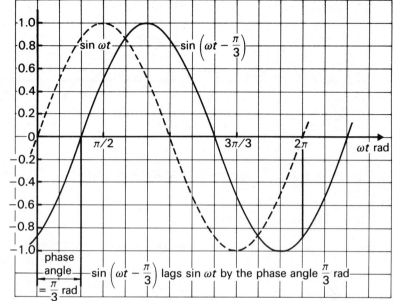

Combining Sine Waves

In the context of the work which follows 'combining' sine waves means *adding sine waves together.* The result of this addition is called the **resultant waveform**. We shall examine the result of adding two sine waves of the same frequency and also of adding two sine waves of different frequencies.

Adding Two Sine Waves of the Same Frequency

The methods used are well illustrated by means of a typical example.

Consider two sinusoidal electric currents represented by the equations $i_1 = 10 \sin \theta°$ and $i_2 = 12 \sin (\theta + 30)°$. If the resultant waveform is denoted by i_r, then

$$i_r = i_1 + i_2 = 10 \sin \theta° + 12 \sin (\theta + 30)°$$

The addition may be achieved by either phasor addition or addition of the sine waves.

Phasor Addition

The addition of phasors is similar to vector addition used when dealing with forces or velocities. The curves of i_1 and i_2 are shown in Fig. 7.9 together with their associated phasors OC′ and OD′. The angle C′OD′ of 30° between these phasors represents the phase angle by which i_2 leads i_1. In order to add phasors OC′ and OD′ we construct the parallelogram OC′B′D′. Then diagonal OB′ gives the resultant phasor. The amplitude or peak value of i_r is given by the length of OB′ which is 21.3, found by measurement. The phase angle between i_r and i_1 is given by the angle C′OB′. Thus i_r leads i_1 by 16°, found by measurement.

Fig. **7.9**

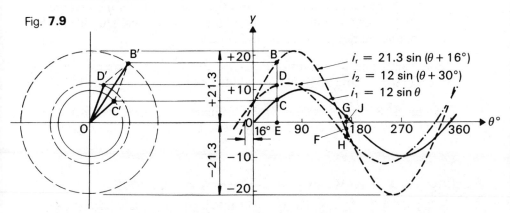

$$i_r = 21.3 \sin (\theta + 16°)$$
$$i_2 = 12 \sin (\theta + 30°)$$
$$i_1 = 12 \sin \theta$$

Addition of the Sine Waves

In order to plot the waveform for i_r we need values of the expression $10 \sin \theta° + 12 \sin (\theta + 30)°$ for suitable values of θ from 0° to 360°. A suitable calculator sequence using, for example, $\theta = 25°$ is as follows, remembering to set first the angle mode on the machine to 'degrees'.

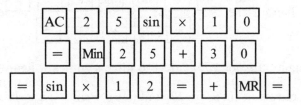

giving an answer 14.1 correct to three significant figures.

The graphs of i_1, i_2 and i_r are shown in Fig. 7.9 where it will be seen that i_r is a sine wave with a peak value of 21.3 and a phase angle of 16°. It has the equation $i_r = 21.3 \sin (\theta + 16°)$.

The peak value of 21.3 and the phase angle of 16° are only approximate as their accuracy depends on reading values from the scales of the graphs. In this case it is possible to calculate these values by a theoretical method and obtain more accurate results. The method is as shown on p. 107 and gives the more accurate answers of 21.254 and 16°24′ which shows the answers we obtained are as good as could be expected from a graphical method.

It is possible to obtain the i_r curve from the graphs of i_1 and i_2 by graphical addition. This is, in fact, similar to adding the values of i_1 and i_2 for various values of the angle θ.

First plot the graphs of i_1 and i_2 and then the points on the i_r curve may be plotted by adding the ordinates (i.e. the vertical lengths) of the corresponding points on the i_1 and i_2 curves.

For example, to plot point B on the i_r curve in Fig. 7.9 we can measure the lengths CE and DE and then add their values, since BE $=$ CE $+$ DE.

We must take care to allow for the ordinates being positive or negative. For example, to find point F on the i_r curve we must use FJ $=$ GJ $-$ HJ.

Adding Two Sine Waves of Different Frequencies

Since we desire to see the shape of the resultant curve, we shall add the given sine waves. Let us see the result of adding together the sine curves $\sin \theta$ and $\sin 2\theta$. Note that $\sin 2\theta$ has a frequency twice that of $\sin \theta$.

In order to plot the resultant curve we need values of the expression $(\sin \theta + \sin 2\theta)$. As usual we shall use values of θ from 0° to 360°. A suitable calculator sequence using, for example, $\theta = 60°$ is as follows:

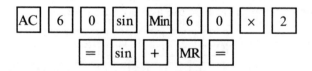

giving an answer 1.73 correct to three significant figures.

The curves of $\sin \theta$, $\sin 2\theta$, and $(\sin \theta + \sin 2\theta)$ are shown plotted in Fig. 7.10.

We can see that the graph of $(\sin \theta + \sin 2\theta)$ is non-sinusoidal (i.e. *not* the shape of a sine curve). However, the curve is a waveform (or a periodic function) since it will repeat indefinitely.

Fig. **7.10**

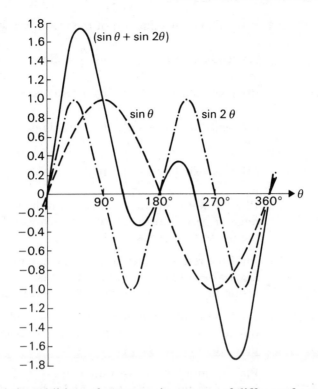

In general the addition of any two sine waves of different frequencies will result in a non-sinusoidal waveform. You may check this by working through the examples in Question 10 of Exercise 7.2 which follows.

EXERCISE

1 Plot the graphs of $\sin \theta$ and $\sin (\theta + 0.9)$ on the same axes on an angular base using units in radians. Indicate the phase angle between the waveforms and explain whether it is an angle of lead or lag.

2 Plot the graph of $\sin \left(\omega t + \dfrac{\pi}{6} \right)$ and $\sin \omega t$ on the same axes on an angular base showing a cycle of each waveform. Identify the phase angle between the curves.

3 Plot the graphs of $\sin \left(\omega t + \dfrac{\pi}{3} \right)$ and $\sin \left(\omega t - \dfrac{\pi}{4} \right)$ on the same axes on an angular base showing a cycle of each waveform. Identify the phase angle between the curves.

4 Write down the equation of the waveform which:

    (a) leads $\sin \omega t$ by $\dfrac{\pi}{2}$ radians.

    (b) lags $\sin \omega t$ by π radians.

    (c) leads $\sin\left(\omega t - \dfrac{\pi}{3}\right)$ by $\dfrac{\pi}{3}$ radians.

    (d) lags $\sin\left(\omega t + \dfrac{\pi}{6}\right)$ by $\dfrac{\pi}{3}$ radians.

5 Plot the waves of $\sin \theta$ and $\cos \theta$ on the same axes on an angular base showing a cycle of each waveform. Identify the phase angle between the curves.

6 Find the resultant waveform of the curves $3 \sin \theta$ and $2 \cos \theta$ by graphical addition. What is the amplitude of the resultant waveform and the phase angle relative to $3 \sin \theta$?

7 Find the resultant voltage v_r of the two voltages represented by the equations $v_1 = 3 \sin \theta$ and $v_2 = 5 \sin (\theta - 30°)$ by plotting the three graphs.

8 Plot the graphs of $i_1 = 5 \sin \theta$ and $i_2 = 2 \sin (\theta + 45°)$ and hence find the resultant of i_1 and i_2 by graphical addition. State the equation of the resultant current i_r.

9 The voltages v_1 and v_2 are represented by the equations $v_1 = 30 \sin (\theta + 60°)$ and $v_2 = 50 \sin (\theta - 45°)$. Plot the curves of v_1 and v_2 and the resultant voltage v_r and find the equation representing v_r.

10 Find the resultant curve in each of the following and verify that the resultant is a non-sinusoidal periodic waveform:

    (a) $\sin \theta + \sin \dfrac{\theta}{2}$ (b) $3 \sin \theta + 2 \sin 2\theta$

    (c) $\sin \dfrac{\theta}{2} + 2 \sin 2\theta$ (d) $\sin (\theta + 30°) + \sin 2\theta$

Trigonometrical Formulae

OBJECTIVES

1 Use the formulae for sin $(A \pm B)$, cos $(A \pm B)$, tan $(A \pm B)$.
2 Derive the double-angle formulae for sin $2A$, cos $2A$, tan $2A$.
3 Derive half-angle formulae for sin A, cos A, tan A.
4 Derive formulae for sin $P \pm$ sin Q and cos $P \pm$ cos Q.
5 Express R sin $(\omega t \pm \alpha)$ in the form

$$a \sin \omega t \pm b \cos \omega t$$

using **1** and vice versa.
6 Deduce the relationship between a, b, R and α.

Introduction

This chapter may appear to be just a collection of trigonometrical formulae with no practical application. However, most of the formulae are used in the solving of problems in technology. They may be thought of as essential tools and you should bear this in mind as you work through this chapter.

Compound Angle Formulae

It can be shown that

$$\sin (A + B) = \sin A \cos B + \cos A \sin B \qquad [1]$$

$$\sin (A - B) = \sin A \cos B - \cos A \sin B \qquad [2]$$

$$\cos (A + B) = \cos A \cos B - \sin A \sin B \qquad [3]$$

$$\cos (A - B) = \cos A \cos B + \sin A \sin B \qquad [4]$$

On inspecting these formulae you may feel that an error has been made in the 'signs' on the right-hand sides of the equations for cos $(A \pm B)$.

They are correct, however, and a special note should be made of them.

Since the above formulae involve the two angles A and B they are called **compound angle** formulae.

The following examples show some uses of the above formulae.

Example 8.1

Simplify:

(a) $\sin(\theta + 90°)$,
(b) $\cos(\theta - 270°)$.

(a) Using $\quad \sin(A + B) = \sin A \cos B + \cos A \sin B$

and substituting θ for A and $90°$ for B,

we have $\quad \sin(\theta + 90°) = \sin \theta \cos 90° + \cos \theta \sin 90°$

$$= (\sin \theta)0 + (\cos \theta)1$$

$$= \cos \theta$$

(b) Using

$$\cos(A - B) = \cos A \cos B + \sin A \sin B$$

and substituting θ for A and $270°$ for B,

we have

$$\cos(\theta - 270°) = \cos \theta \cos 270° + \sin \theta \sin 270°$$

$$= (\cos \theta)0 + (\sin \theta)(-1)$$

$$= -\sin \theta$$

tan $(A + B)$ Formula

Now if we divide Equation [1] by Equation [3] we have

$$\frac{\sin(A + B)}{\cos(A + B)} = \frac{\sin A \cos B + \cos A \sin B}{\cos A \cos B - \sin A \sin B}$$

and if we divide both numerator and denominator of the right-hand side by $\cos A \cos B$,

then
$$\tan (A + B) = \dfrac{\dfrac{\sin A \cos B}{\cos A \cos B} + \dfrac{\cos A \sin B}{\cos A \cos B}}{\dfrac{\cos A \cos B}{\cos A \cos B} - \dfrac{\sin A \sin B}{\cos A \cos B}}$$

∴
$$\boxed{\tan (A + B) = \dfrac{\tan A + \tan B}{1 - \tan A \tan B}}$$
[5]

Example 8.2

By using the sines, cosines and tangents of 30° and 45° established from suitable right-angled triangles find:

(a) sin 75°
(b) cos 15°
(c) tan 75°

To find the trigonometrical ratios of 30° and 45° we use suitable right-angled triangles shown in Figs 8.1 and 8.2.

Fig. **8.1**

Fig. **8.2**

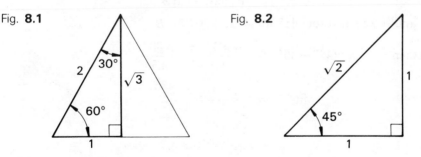

From Fig. 8.1 we have
$$\sin 30° = \frac{1}{2}, \quad \cos 30° = \frac{\sqrt{3}}{2} \quad \text{and} \quad \tan 30° = \frac{1}{\sqrt{3}}$$

and from Fig. 8.2 we have
$$\sin 45° = \frac{1}{\sqrt{2}}, \quad \cos 45° = \frac{1}{\sqrt{2}} \quad \text{and} \quad \tan 45° = 1$$

(a) Now
$$\sin (45° + 30°) = \sin 45° \cos 30° + \cos 45° \sin 30°$$

$$\therefore \qquad \sin 75° = \left(\frac{1}{\sqrt{2}}\right)\left(\frac{\sqrt{3}}{2}\right) + \left(\frac{1}{\sqrt{2}}\right)\left(\frac{1}{2}\right)$$

$$= \frac{\sqrt{3}+1}{2\sqrt{2}}$$

$$= 0.966$$

(b) Now $\cos(A - B) = \cos A \cos B + \sin A \sin B$

and if we substitute 45° for A and 30° for B

then $\cos(45° - 30°) = \cos 45° \cos 30° + \sin 45° \sin 30°$

$$\therefore \qquad \cos 15° = \left(\frac{1}{\sqrt{2}}\right)\left(\frac{\sqrt{3}}{2}\right) + \left(\frac{1}{\sqrt{2}}\right)\left(\frac{1}{2}\right)$$

$$= \frac{\sqrt{3}+1}{2\sqrt{2}}$$

$$= 0.966$$

This confirms the fact that

$$\sin 75° = \cos 15°$$

(c) Now $\tan(A + B) = \dfrac{\tan A + \tan B}{1 - \tan A \tan B}$

and if we substitute 45° for A, and 30° for B

then $\tan(45° + 30°) = \dfrac{\tan 45° + \tan 30°}{1 - \tan 45° \tan 30°}$

$$\therefore \qquad \tan 75° = \frac{1 + 1/\sqrt{3}}{1 - 1(1/\sqrt{3})}$$

$$= 3.732$$

Example 8.3

Find the angle between 0° and 90° which satisfies the equation

$$\sin x = 2 \sin(x - 20°)$$

We have $\sin x = 2 \sin(x - 20°)$

$$= 2[\sin x \cos 20° - \cos x \sin 20°]$$

$$= 2[(\sin x)0.9397 - (\cos x)0.3420]$$

$$= 1.8794(\sin x) - 0.6840(\cos x)$$

$$\therefore \qquad 0.6840 \cos x = 1.8794 \sin x - \sin x$$

$$\therefore \qquad 0.6840 \cos x = 0.8794 \sin x$$

$$\therefore \qquad \frac{\sin x}{\cos x} = \frac{0.6840}{0.8794}$$

$$\therefore \qquad \tan x = 0.7778$$

$$\therefore \qquad x = 37.88°$$

Double-angle Formulae

The term 'double angle' refers to angle $2A$. A double-angle formula expresses a trigonometrical ratio of a double angle in terms of trigonometrical ratio(s) of the single angle.

sin 2A Formula

If we put $B = A$ into Equation [1] then we have

$$\sin (A + A) = \sin A \cos A + \cos A \sin A$$

or
$$\boxed{\sin 2A = 2 \sin A \cos A} \qquad\qquad [6]$$

cos 2A Formulae

If we put $B = A$ into Equation [3], then we have

$$\cos (A + A) = \cos A \cos A - \sin A \sin A$$

or
$$\boxed{\cos 2A = \cos^2 A - \sin^2 A} \qquad\qquad [7]$$

For any angle we know that $\sin^2 A + \cos^2 A = 1$

from which either $\qquad\qquad\qquad\qquad \sin^2 A = 1 - \cos^2 A$

or $\qquad\qquad\qquad\qquad\qquad\qquad \cos^2 A = 1 - \sin^2 A$

Now substituting $\sin^2 A = 1 - \cos^2 A$ into Equation [7], we have

$$\cos 2A = \cos^2 A - (1 - \cos^2 A)$$

or
$$\boxed{\cos 2A = 2 \cos^2 A - 1} \qquad\qquad [8]$$

Also substituting $\cos^2 A = 1 - \sin^2 A$ into Equation [7], we have

$$\cos 2A = (1 - \sin^2 A) - \sin^2 A$$

or

$$\boxed{\cos 2A = 1 - 2\sin^2 A} \qquad [9]$$

tan 2A formula

If we put $B = A$ into Equation [5], then we have

$$\tan(A + A) = \frac{\tan A + \tan A}{1 - \tan A \tan A}$$

or

$$\boxed{\tan 2A = \frac{2\tan A}{1 - \tan^2 A}} \qquad [10]$$

Example 8.4

If $\cos \phi = \frac{24}{25}$ find, without calculating the value of the angle ϕ, the values of:

(a) $\sin 2\phi$ (b) $\cos 2\phi$ (c) $\tan 2\phi$

Using the right-angled triangle ABC shown in Fig. 8.3 and the theorem of Pythagoras,

Fig. **8.3**

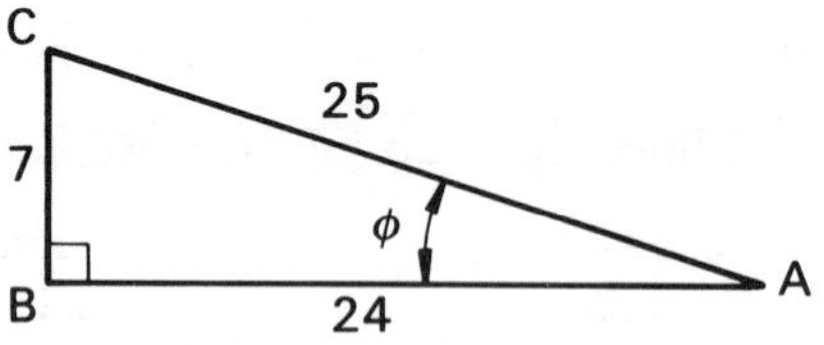

we have $CB^2 = AC^2 - BA^2 = 25^2 - 24^2 = 49$

Thus $CB = 7$

Hence $\sin \phi = \frac{7}{25}$, $\cos \phi = \frac{24}{25}$ and $\tan \phi = \frac{7}{24}$

(a) Now $\sin 2\phi = 2\sin \phi \cos \phi = 2(\frac{7}{25})(\frac{24}{25}) = 0.538$

(b) Now $\cos 2\phi = 2\cos^2 \phi - 1 = 2(\frac{24}{25})^2 - 1 = 0.843$

(c) Now $\tan 2\phi = \dfrac{2\tan \phi}{1 - \tan^2 \phi} = \dfrac{2(7/24)}{1 - (7/24)^2} = 0.638$

Half-angle Formulae

The term 'half angle' refers to angle $\frac{A}{2}$. A half-angle formula expresses a trigonometrical ratio of a single angle A in terms of the tangent of the half-angle $\frac{A}{2}$.

We know that $\qquad\qquad \sin^2 B + \cos^2 B = 1$

and so Equation [6] may be written as

$$\sin 2B = \frac{2 \sin B \cos B}{\cos^2 B + \sin^2 B}$$

and if we divide both the numerator and denominator of the right-hand side by $\cos^2 B$,

we have $\qquad\qquad \sin 2B = \dfrac{\dfrac{2 \sin B \cos B}{\cos^2 B}}{\dfrac{\cos^2 B}{\cos^2 B} + \dfrac{\sin^2 B}{\cos^2 B}}$

$\therefore \qquad\qquad \sin 2B = \dfrac{2 \tan B}{1 + \tan^2 B}$

Now if we put $B = \frac{A}{2}$ then we obtain the half-angle formula

$$\sin A = \frac{2 \tan \dfrac{A}{2}}{1 + \tan^2 \dfrac{A}{2}} \qquad\qquad [11]$$

Now if we use Equation [7] and use a method similar to the above we can obtain the half-angle formula

$$\cos A = \frac{1 - \tan^2 \dfrac{A}{2}}{1 + \tan^2 \dfrac{A}{2}} \qquad\qquad [12]$$

Finally, from Equation [10] we have the half-angle formula

$$\tan A = \frac{2 \tan \dfrac{A}{2}}{1 - \tan^2 \dfrac{A}{2}} \qquad [13]$$

Sum or Difference of Trigonometrical Ratios as Products

Suppose we put $\qquad P = A + B$

and $\qquad Q = A - B$

Then adding we have $\qquad P + Q = 2A$

and subtracting we have $\qquad P - Q = 2B$

$\therefore \qquad A = \dfrac{P+Q}{2} \quad$ and $\quad B = \dfrac{P-Q}{2}$

Now Equation [1] states $\qquad \sin(A + B) = \sin A \cos B + \cos A \sin B$

and Equation [2] states $\qquad \sin(A - B) = \sin A \cos B - \cos A \sin B$

Adding, $\qquad \sin(A + B) + \sin(A - B) = 2 \sin A \cos B$

And introducing P and Q from the above expressions we get

$$\sin P + \sin Q = 2 \sin\left(\frac{P+Q}{2}\right) \cos\left(\frac{P-Q}{2}\right) \qquad [14]$$

Also by subtracting Equation [2] from Equation [1] we get

$$\sin(A + B) - \sin(A - B) = 2 \cos A \sin B$$

And introducing P and Q we get

$$\sin P - \sin Q = 2 \cos\left(\frac{P+Q}{2}\right) \sin\left(\frac{P-Q}{2}\right) \qquad [15]$$

You may now like to follow a similar method using Equations [3] and [4] and obtain the equations

$$\cos P + \cos Q = 2 \cos\left(\frac{P+Q}{2}\right) \cos\left(\frac{P-Q}{2}\right) \qquad [16]$$

and

$$\cos P - \cos Q = -2 \sin\left(\frac{P+Q}{2}\right) \sin\left(\frac{P-Q}{2}\right) \qquad [17]$$

Example 8.5

Show that $\sin\theta + \sin 3\theta - \sin 5\theta - \sin 7\theta = -4\sin 2\theta \cos\theta \cos 4\theta$

Now left-hand side (LHS) $= (\sin\theta + \sin 3\theta) - (\sin 5\theta + \sin 7\theta)$

and using Equation [14] twice we get

$$\text{LHS} = 2\sin\left(\frac{\theta + 3\theta}{2}\right)\cos\left(\frac{\theta - 3\theta}{2}\right) - 2\sin\left(\frac{5\theta + 7\theta}{2}\right)\cos\left(\frac{5\theta - 7\theta}{2}\right)$$

$$= 2\sin 2\theta \cos(-\theta) - 2\sin 6\theta \cos(-\theta)$$

$$= 2\cos(-\theta)\,[\sin 2\theta - \sin 6\theta]$$

Now since $\cos(-\theta) = \cos\theta$, using Equation [15] gives

$$\text{LHS} = 2\cos\theta\left[2\cos\left(\frac{2\theta + 6\theta}{2}\right)\sin\left(\frac{2\theta - 6\theta}{2}\right)\right]$$

$$= 2\cos\theta\,[2\cos 4\theta \sin(-2\theta)]$$

and knowing that $\sin(-2\theta) = -\sin 2\theta$, we have

$$\text{LHS} = -4\sin 2\theta \cos\theta \cos 4\theta$$

$$= \text{right-hand side.}$$

EXERCISE

1 Simplify:
   (a) $\sin(x + 180°)$ (b) $\cos(180° - x)$
   (c) $\sin(90° - x)$ (d) $\cos(270° + x)$
   (e) $\tan(x + 180°)$

2 Find the value of $\sin 15°$ using the fact
   $\sin 15° = \sin(45° - 30°)$.

3 Show that $\sin(\phi + 60°) + \sin(\phi - 60°) = \sin\phi$.

4 If $\tan\theta = \frac{3}{4}$ and $\tan\phi = \frac{5}{12}$ sketch suitable right-angled
   triangles from which the other trigonometrical ratios of θ
   and ϕ may be found. Hence find the values of:
   (a) $\sin(\theta - \phi)$ (b) $\cos(\theta + \phi)$ (c) $\tan(\theta + \phi)$
   (d) $\sin 2\theta$ (e) $\cos 2\phi$ (f) $\tan 2\theta$

5 Show that $\sqrt{2}\sin\left(\theta - \frac{\pi}{4}\right) = \sin\theta - \cos\theta$.

6 Find the angle between $0°$ and $90°$ that satisfies the equation $\cos\theta = 3\cos(\theta + 30°)$.

7 Illustrate the facts that $\cos\alpha = \frac{3}{5}$ and $\cos\beta = \frac{4}{5}$ using a right-angled triangle. Without calculating the values of the angles α and β,

(a) show that $\alpha + \beta = 90°$
(b) find the values of $\sin 2\alpha$, $\cos 2\beta$, and $\tan 2\alpha$.

8 Find the angle between $0°$ and $90°$ that satisfies the equation $2\cos(\theta + 60°) = \sin(\theta + 30°)$.

9 Find the angle between $0°$ and $90°$ that satisfies the equation $(1 + \tan A)\tan 2A = 1$.

10 If $\sin 33°\,24' = 0.5505$ find, using the compound angle formulae, the values of:

(a) $\cos 236°\,36'$ (b) $\sin 326°\,36'$

11 Show that
$$\cos 2\theta + \cos 4\theta + \cos 6\theta + \cos 12\theta = 4\cos 3\theta \cos 4\theta \cos 5\theta.$$

12 Show that $\dfrac{\sin\alpha + \sin\beta}{\cos\alpha + \cos\beta} = \tan\left(\dfrac{\alpha + \beta}{2}\right)$.

13 Show that $\dfrac{\sin\phi + \sin 2\phi}{\cos\phi - \cos 2\phi} = \cot\dfrac{\phi}{2}$.

The Form $R \sin(\theta \pm \alpha)$

Consider $3\sin(\theta + 40°)$.

Now $\sin(A + B) = \sin A \cos B + \cos A \sin B$

$\therefore$ $3\sin(\theta + 40°) = 3[\sin\theta \cos 40° + \cos\theta \sin 40°]$

$= 3[(\sin\theta)0.7660 + (\cos\theta)0.6428]$

$= 2.298\sin\theta + 1.928\cos\theta$

Now using the same method, if R and α are constants,

then $R\sin(\theta \pm \alpha) = R[\sin\theta \cos\alpha \pm \cos\theta \sin\alpha]$

$= (R\cos\alpha)\sin\theta \pm (R\sin\alpha)\cos\theta$

$= a\sin\theta \pm b\cos\theta$

where $$a = R \cos \alpha \qquad [18]$$

and $$b = R \sin \alpha \qquad \cdot \qquad [19]$$

∴ squaring and adding Equations [18] and [19] gives

$$(R \cos \alpha)^2 + (R \sin \alpha)^2 = a^2 + b^2$$

∴ $$R^2 \cos^2\alpha + R^2 \sin^2\alpha = a^2 + b^2$$

∴ $$R^2(\cos^2\alpha + \sin^2\alpha) = a^2 + b^2$$

and since $\cos^2\alpha + \sin^2\alpha = 1$

$$R^2 = a^2 + b^2$$

Also dividing Equation [19] by equation [18] gives

$$\frac{R \sin \alpha}{R \cos \alpha} = \frac{b}{a}$$

∴ $$\tan \alpha = \frac{b}{a}$$

Hence
$$R \sin (\theta \pm \alpha) = a \sin \theta \pm b \cos \theta$$

where
$$R^2 = a^2 + b^2 \quad \text{and} \quad \tan \alpha = \frac{b}{a}$$

Problems using the above relationship usually occur in the reverse order to the sequence above.

Example 8.6

Express $4 \sin \theta - 3 \cos \theta$ in the form $R \sin (\theta - \alpha)$.

Comparing the given expression with $a \sin \theta - b \cos \theta$ we have $a = 4$ and $b = 3$,

and since $$R^2 = a^2 + b^2$$

$$R^2 = 4^2 + 3^2$$

∴ $$R = \sqrt{16 + 9}$$

∴ $$R = 5$$

Also $\tan \alpha = \dfrac{b}{a}$

and $\tan \alpha = \dfrac{3}{4}$

                                $= 0.75$

$\therefore$ $\alpha = 36.87°$

Hence $4 \sin \theta - 3 \cos \theta = 5 \sin (\theta - 36.87°)$

Example 8.7

Express $6 \sin \omega t - 8 \cos \omega t$ in the form $R \sin (\theta - \alpha)$. Hence find the maximum value of $6 \sin \omega t - 8 \cos \omega t$ and the value of ωt at which it occurs.

This example is similar to Example 8.6 except that the angles ωt and α are in radians.

Comparing the given expression with $a \sin \theta + b \cos \theta$ we have $a = 6$ and $b = 8$.

Then $R = \sqrt{6^2 + 8^2} = 10$

and $\tan \alpha = \dfrac{8}{6} = 1.333$

$\therefore$ $\alpha = 0.927 \text{ rad}$

Hence $6 \sin \omega t - 8 \cos \omega t = 10 \sin (\omega t - 0.927)$

Now the maximum value of the sine of an angle is unity, and occurs when the angle is $90°$ or $\dfrac{\pi}{2}$ radians.

Thus $6 \sin \omega t - 8 \cos \omega t$ will be a maximum

when $\sin (\omega t - 0.927) = 1$

i.e. when $\omega t - 0.927 = \dfrac{\pi}{2} = 1.571$

or $\omega t = 1.571 + 0.927 = 2.498$

Therefore, the maximum value of $6 \sin \omega t - 8 \cos \omega t$ is 10 and it occurs when $\omega t = 2.498$ rad.

Example 8.8

Express $10 \sin \theta + 12 \sin (\theta + 30)°$ in the form $R \sin (\theta + \alpha)°$.

Using the expression

$$\sin (A + B) = \sin A \cos B + \cos A \sin B$$

we have

$$10 \sin \theta + 12 \sin (\theta + 30)° = 10 \sin \theta + 12 (\sin \theta \cos 30° \\ + \cos \theta \sin 30°)$$

$$= 10 \sin \theta + 12 \cos 30° \sin \theta \\ + 12 \sin 30° \cos \theta$$

$$= 10 \sin \theta + 10.4 \sin \theta + 6 \cos \theta$$

$$= 20.4 \sin \theta + 6 \cos \theta$$

Now comparing this expression with $a \sin \theta + b \cos \theta$ we have

$$R^2 = 20.4^2 + 6^2 \quad \text{and} \quad \tan \alpha = \frac{6}{20.4}$$

$$\therefore \qquad R = 21.3 \qquad \text{and} \qquad \alpha = 16.3°$$

Thus $10 \sin \theta + 12 \sin (\theta + 30)° = 21.3 \sin (\theta + 16.3)°$

This result confirms the answer obtained graphically on p. 91.

EXERCISE

8·2

1 Express $3 \sin \theta + 2 \cos \theta$ in the form $R \sin (\theta + \alpha)$.

2 Express $7 \cos \omega t + \sin \omega t$ in the form $R \sin (\omega t + \alpha)$.

3 Rewrite $5 \sin \omega t - 7 \cos \omega t$ in the form $R \sin (\omega t - \alpha)$.

4 Using the result obtained in Question 3 find the maximum value of $5 \sin \omega t - 7 \cos \omega t$ and the value of ωt at which this occurs.

5 An electric current is given by $i = 200 \sin 300t + 100 \cos 300t$. Express this as a single trigonometrical function and find its maximum value.

6 Express $3 \sin \theta + 5 \sin (\theta - 30°)$ in the form $R \sin (\theta + \alpha)$.

7 Express $5 \sin \theta + 2 \sin (\theta + 45°)$ in the form $R \sin (\theta + \alpha)$.

8 Express $30 \sin (\theta + 60°) + 50 \sin (\theta - 45°)$ in the form $R \sin (\theta \pm \alpha)$.

Complex Numbers

OBJECTIVES

1 Understand the necessity of extending the number system to include the square roots of negative numbers.

2 Define j as $\sqrt{-1}$.

3 Define a complex number as consisting of a real part and an imaginary part.

4 Define a complex number z in the algebraic form $x + jy$.

5 Determine the complex roots of $ax^2 + bx + c = 0$ when $b^2 < 4ac$ using the quadratic formula.

6 Perform the addition and subtraction of complex numbers in algebraic form.

7 Define the conjugate of a complex number in algebraic form.

8 Perform the multiplication and division of complex numbers in algebraic form.

9 Represent the algebraic form of a complex number on an Argand diagram, and show how it may be represented as a phasor.

10 Deduce that j may be considered to be an operator, such that when the phasor representing $x + jy$ is multiplied by j it rotates the phasor through 90° anticlockwise.

11 Understand how phasors on an Argand diagram may be added and subtracted in a manner similar to the addition and subtraction of vectors.

12 Show that the full polar form of a complex number is $(\cos\theta + j\sin\theta)$, which may be abbreviated to $r\underline{/\theta}$.

13 Perform the operations involved in the conversion of complex numbers in algebraic form to polar form and vice versa.

14 Multiply and divide numbers in polar form.

15 Determine the square roots of a complex number.

16 Apply the above to problems arising from relevant engineering technology.

Introduction

The solution of the quadratic equation $ax^2 + bx + c = 0$ is given by the formula

$$x = \frac{-b \pm \sqrt{b^2 - 4ac}}{2a}$$

When we use this formula most of the quadratic equations we meet, when solving technology problems, are found to have roots which are ordinary positive or negative numbers.

Consider now the equation $x^2 - 4x + 13 = 0$.

Then
$$x = \frac{-(-4) \pm \sqrt{(-4)^2 - 4 \times 1 \times 13}}{2 \times 1}$$

$$= \frac{4 \pm \sqrt{16 - 52}}{2}$$

$$= \frac{4 \pm \sqrt{-36}}{2}$$

$$= \frac{4 \pm \sqrt{(-1)(36)}}{2}$$

$$= \frac{4 \pm \sqrt{(-1)} \times \sqrt{(36)}}{2}$$

$$= \frac{4}{2} \pm \sqrt{-1} \times \frac{6}{2}$$

$$= 2 \pm \sqrt{-1} \times 3$$

It is not possible to find the value of the square root of a negative number.

In order to try to find a meaning for roots of this type we represent $\sqrt{-1}$ by the symbol j.

(Books on pure mathematics often use the symbol i, but in technology j is preferred as i is used for the instantaneous value of a current.)

Thus the roots of the above equation become $2 + j3$ and $2 - j3$.

Definitions

Expressions such as $2 + j3$ are called **complex numbers**. The number 2 is called the **real part** and j3 is called the **imaginary part**.

The general expression for a complex number is $x + jy$, which has a real part equal to x and an imaginary part equal to jy. The form $x + jy$ is said to be the **algebraic form** of a complex number. It may also be called the **Cartesian form** or **rectangular notation**.

Powers of j

We have defined j such that

$$j = \sqrt{-1}$$

∴ squaring both sides of the equation gives

$$j^2 = (\sqrt{-1})^2 = -1$$

Hence $j^3 = j^2 \times j = (-1) \times j = -j$

and $j^4 = (j^2)^2 = (-1)^2 = 1$

and $j^5 = j^4 \times j = 1 \times j = j$

and $j^6 = (j^2)^3 = (-1)^3 = -1$

and so on.

The most used of the above relationships is $j^2 = -1$.

Addition and Subtraction of Complex Numbers in Algebraic Form

The real and imaginary parts must be treated separately. The real parts may be added and subtracted and also the imaginary parts may be added and subtracted, both obeying the ordinary laws of algebra.

Thus
$$(3 + j2) + (5 + j6) = 3 + j2 + 5 + j6$$
$$= (3 + 5) + j(2 + 6)$$
$$= 8 + j8$$
and
$$(1 - j2) - (-4 + j) = 1 - j2 + 4 - j$$
$$= (1 + 4) - j(2 + 1)$$
$$= 5 - j3$$

Example 9.1

If z_1, z_2 and z_3 represent three complex numbers such that $z_1 = 1.6 + j2.3$, $z_2 = 4.3 - j0.6$ and $z_3 = -1.1 - j0.9$ find the complex numbers which represent:

(a) $z_1 + z_2 + z_3$,

(b) $z_1 - z_2 - z_3$.

(a) $z_1 + z_2 + z_3 = (1.6 + j2.3) + (4.3 - j0.6) + (-1.1 - j0.9)$

$= 1.6 + j2.3 + 4.3 - j0.6 - 1.1 - j0.9$

$= (1.6 + 4.3 - 1.1) + j(2.3 - 0.6 - 0.9)$

$= 4.8 + j0.8$

(b) $z_1 - z_2 - z_3 = (1.6 + j2.3) - (4.3 - j0.6) - (-1.1 - j0.9)$

$= 1.6 + j2.3 - 4.3 + j0.6 + 1.1 + j0.9$

$= (1.6 - 4.3 + 1.1) + j(2.3 + 0.6 + 0.9)$

$= -1.6 + j3.8$

Multiplication of Complex Numbers in Algebraic Form

Consider the product of two complex numbers, $(3 + j2)(4 + j)$.

The brackets are treated in exactly the same way as they are in ordinary algebra, such that

$$(a + b)(c + d) = ac + bc + ad + bd$$

Hence $(3 + j2)(4 + j) = 3 \times 4 + j2 \times 4 + 3 \times j + j2 \times j$

$= 12 + j8 + j3 + j^2 2$

$= 12 + j8 + j3 - 2 \qquad$ since $\quad j^2 = -1$

$= (12 - 2) + j(8 + 3)$

$= 10 + j11$

Example 9.2

Express the product of $2 + j$, $-3 + j2$, and $1 - j$ as a single complex number.

Then $(2 + j)(-3 + j2)(1 - j) = (2 + j)(-3 + j2 + j3 - j^2 2)$

$= (2 + j)(-1 + j5) \quad$ since $\quad j^2 = -1$

$= -2 - j + j10 + j^2 5$

$= -7 + j9 \qquad$ since $\quad j^2 = -1$

Conjugate Complex Numbers

Consider
$$(x + jy)(x - jy) = x^2 + jxy - jxy - j^2y$$
$$= x^2 - (-1)y^2$$
$$= x^2 + y^2$$

Hence we have the product of two complex numbers which produces a real number since it does not have a j term. If $x + jy$ represents a complex number then $x - jy$ is known as its **conjugate** (and vice versa). For example, the conjugate of $(3 + j4)$ is $(3 - j4)$ and their product is

$$(3 + j4)(3 - j4) = 9 + j12 - j12 - j^2 16 = 9 - (-1)16$$
$$= 25 \quad \text{which is a real number}$$

Division of Complex Numbers in Algebraic Form

Consider $\dfrac{(4 + j5)}{(1 - j)}$. We use the method of rationalising the denominator.

This means removing the j terms from the bottom line of the fraction. If we multiply $(1 - j)$ by its conjugate $(1 + j)$ the result will be a real number. Hence, in order not to alter the value of the given expression, we multiply both the numerator and the denominator by $(1 + j)$.

Thus
$$\frac{(4 + j5)}{(1 - j)} = \frac{(4 + j5)(1 + j)}{(1 - j)(1 + j)}$$
$$= \frac{4 + j5 + j4 + j^2 5}{1 - j + j - j^2}$$
$$= \frac{4 + j9 + (-1)5}{1 - (-1)}$$
$$= \frac{-1 + j9}{2}$$
$$= -\frac{1}{2} + j\frac{9}{2}$$
$$= -0.5 + j4.5$$

Example 9.3

The impedance Z of a circuit having a resistance and inductive reactance in series is given by the complex number $Z = 5 + j6$.

Find the admittance Y of the circuit if $Y = \dfrac{1}{Z}$.

Now
$$Y = \frac{1}{Z} = \frac{1}{5 + j6}$$

The conjugate of the denominator is $5 - j6$ and we therefore multiply both the numerator and denominator by $5 - j6$.

Then

$$Y = \frac{(5 - j6)}{(5 + j6)(5 - j6)}$$

$$= \frac{5 - j6}{25 + j30 - j30 - j^2 36}$$

$$= \frac{5 - j6}{25 - (-1)36} = \frac{5 - j6}{61} = \frac{5}{61} - j\frac{6}{61} = 0.082 - j0.098$$

Example 9.4

Two impedances Z_1 and Z_2 are given by the complex numbers $Z_1 = 1 + j5$ and $Z_2 = j7$. Find the equivalent impedance Z if

(a) $Z = Z_1 + Z_2$ where Z_1 and Z_2 are in series,

(b) $\dfrac{1}{Z} = \dfrac{1}{Z_1} + \dfrac{1}{Z_2}$ when Z_1 and Z_2 are in parallel.

(a)
$$Z = Z_1 + Z_2 = (1 + j5) + j7$$
$$= 1 + j5 + j7$$
$$= 1 + j12$$

(b)
$$\frac{1}{Z} = \frac{1}{Z_1} + \frac{1}{Z_2} = \frac{1}{(1+j5)} + \frac{1}{j7}$$

$$= \frac{j7 + (1+j5)}{(1+j5)j7}$$

$$= \frac{1+j12}{j7 + j^2 35}$$

$$= \frac{1+j12}{j7 + (-1)35}$$

Thus
$$Z = \frac{j7 - 35}{1 + j12}$$

$$= \frac{(j7 - 35)(1 - j12)}{(1 + j12)(1 - j12)}$$

$$= \frac{j7 - 35 - j^2 84 + j420}{1 + j12 - j12 - j^2 144}$$

$$= \frac{j427 - 35 - (-1)84}{1 - (-1)144}$$

$$= \frac{49 + j427}{145}$$

$$= 0.338 + j2.945$$

EXERCISE

9·1

1 Add the following complex numbers:
(a) $3 + j5$, $7 + j3$ and $8 + j2$
(b) $2 - j7$, $3 + j8$ and $-5 - j2$
(c) $4 - j2$, $7 + j3$, $-5 - j6$ and $2 - j5$

2 Subtract the following complex numbers:
(a) $3 + j5$ from $2 + j8$
(b) $7 - j6$ from $3 - j9$
(c) $-3 - j5$ from $7 - j8$

3 Simplify the following expressions giving the answers in the form $x + jy$:
(a) $(3 + j3)(2 + j5)$ (b) $(2 - j6)(3 - j7)$
(c) $(4 + j5)^2$ (d) $(5 + j3)(5 - j3)$
(e) $(-5 - j2)(5 + j2)$ (f) $(3 - j5)(3 - j3)(1 + j)$

(g) $\dfrac{1}{2+j5}$ (h) $\dfrac{2+j5}{2-j5}$

(i) $\dfrac{-2-j3}{5-j2}$ (j) $\dfrac{7+j3}{8-j3}$

(k) $\dfrac{(1+j2)(2-j)}{(1+j)}$ (l) $\dfrac{4+j2}{(2+j)(1-j3)}$

4 Find the real and imaginary parts of:

(a) $1+\dfrac{j}{2}$

(b) $j3+\dfrac{2}{j3}$

(c) $(j2)^2 + 3(j)^5 - j(j)$

5 Solve the following equations giving the answers in the form $x+jy$:

(a) $x^2 + 2x + 2 = 0$
(b) $x^2 + 9 = 0$

6 Find the admittance Y of a circuit if $Y = \dfrac{1}{Z}$ where

$Z = 1.3 + j0.6$.

7 Three impedances Z_1, Z_2 and Z_3 are represented by the complex numbers $Z_1 = 2+j$, $Z_2 = 1+j$ and $Z_3 = j2$. Find the equivalent impedance Z if:

(a) $Z = Z_1 + Z_2 + Z_3$

(b) $\dfrac{1}{Z} = \dfrac{1}{Z_1} + \dfrac{1}{Z_2} + \dfrac{1}{Z_3}$

(c) $Z = \dfrac{1}{\dfrac{1}{Z_1} + \dfrac{1}{Z_2}} + Z_3$

The Argand Diagram

When plotting a graph, Cartesian coordinates are generally used to plot the points. Thus the position of the point P (Fig. 9.1) is defined by the coordinates (3, 2) meaning that $x = 3$ and $y = 2$.

Complex numbers may be represented in a similar way on the Argand diagram. The real part of the complex number is plotted along the horizontal real-axis whilst the imaginary part is plotted along the vertical imaginary, or j-axis.

However, a complex number is denoted not by a point but as a **phasor**. A phasor is a line in which regard is paid both to its magnitude and to its direction. Hence in Fig. 9.2 the complex number $4 + j3$ is represented by the phasor $\overrightarrow{OQ}$, the end Q of the line being found by plotting 4 units along the real-axis and 3 units along the j-axis.

Fig. **9.1**

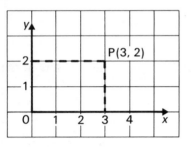

Fig. **9.2**

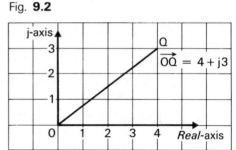

A single letter, the favourite being z, is often used to denote a phasor which represents a complex number. Thus if $z = x + jy$ it is understood that z represents a phasor and not a simple numerical value.

Four typical complex numbers z_1, z_2, z_3 and z_4 are shown on the Argand diagram in Fig. 9.3.

Fig. **9.3**

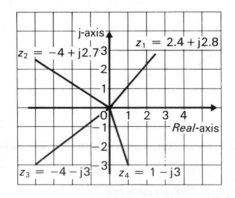

A real number such as 2.7 may be regarded as a complex number with a zero imaginary part, i.e. $2.7 + j0$, and may be represented on the Argand diagram (Fig. 9.4) as the phasor $z = 2.7$ denoted by $\overrightarrow{OA}$ in the diagram.

A number such as j3 is said to be wholly imaginary and may be regarded as a complex number having a zero real part, i.e. $0 + j3$, and may be represented on the Argand diagram (Fig. 9.4) as the phasor $z = j3$ denoted by $\overrightarrow{OB}$ in the diagram.

Fig. **9.4**

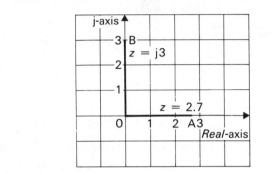

The j-operator

Consider the real number 3 shown on the Argand diagram, in Fig. 9.5.

Fig. **9.5**

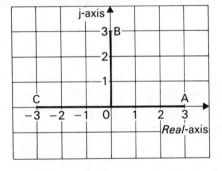

It may be denoted by $\overrightarrow{OA}$. (This is a phasor because it has magnitude and direction.)

If we now multiply the real number 3 by j we obtain the complex number j3 which may be represented by the phasor $\overrightarrow{OB}$.

It follows that the effect of j on phasor $\overrightarrow{OA}$ is to make it become phasor $\overrightarrow{OB}$,

that is $$\overrightarrow{OB} = j\overrightarrow{OA}$$

Hence j is known as an **operator** (called the **j-operator**) which, when applied to a phasor, alters its direction by 90° in an anticlockwise direction without changing its magnitude.

If we now operate on the phasor $\overrightarrow{OB}$ we shall obtain, therefore, phasor $\overrightarrow{OC}$.

In equation form this is

$$OC = j\overrightarrow{OB}$$

but since $\overrightarrow{OB} = j\overrightarrow{OA}$, then

$$\overrightarrow{OC} = j(j\overrightarrow{OA})$$
$$= j^2\overrightarrow{OA}$$
$$= -\overrightarrow{OA} \quad \text{since } j^2 = -1$$

This is true since it may be seen from the phasor diagram that phasor $\overrightarrow{OC}$ is equal in magnitude, but opposite in direction, to phasor $\overrightarrow{OA}$.

Consider now the effect of the j-operator on the complex number $5 + j3$.

In the equation from this is

$$j(5 + j3) = j5 + j(j3)$$
$$= j5 + j^2 3$$
$$= j5 + (-1)3$$
$$= -3 + j5$$

If phasor $z_1 = 5 + j3$ and phasor $z_2 = -3 + j5$, it may be seen from the Argand diagram in Fig. 9.6 that their magnitudes are the same but the effect of the operator j on z_1 has been to alter its direction by 90° anticlockwise to give phasor z_2.

Fig. **9.6**

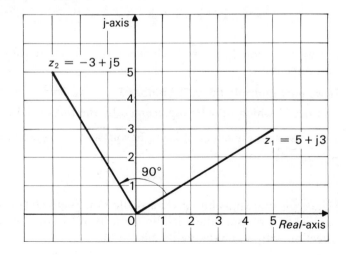

Addition of Phasors

Consider the addition of the two complex numbers $2+j3$ and $4+j2$.

We have
$$(2+j3)+(4+j2) = 2+j3+4+j2$$
$$= (2+4)+j(3+2)$$
$$= 6+j5$$

On the Argand diagram shown in Fig. 9.7, the complex number $2+j3$ is represented by the phasor $\overrightarrow{OA}$, whilst $4+j2$ is represented by phasor $\overrightarrow{OB}$. The addition of the real parts is performed along the real-axis and the addition of the imaginary parts is carried out on the j-axis.

Fig. **9.7**

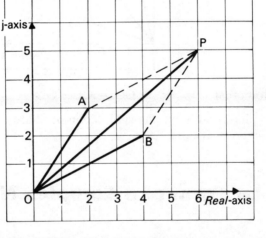

Hence the complex number $6+j5$ is represented by the phasor $\overrightarrow{OP}$.

It follows that
$$\overrightarrow{OP} = \overrightarrow{OB} + \overrightarrow{OA}$$

Hence the addition of phasors is similar to vector addition used when dealing with forces or velocities.

Subtraction of Phasors

Consider the difference of the two complex numbers, $4+j5$ and $1+j4$.

We have
$$(4+j5)-(1+j4) = 4+j5-1-j4$$
$$= (4-1)+j(5-4)$$
$$= 3+j$$

On the Argand diagram shown in Fig. 9.8, the complex number $4 + j5$ is represented by the phasor $\overrightarrow{OC}$, whilst $1 + j4$ is represented by the phasor $\overrightarrow{OD}$. The subtraction of the real parts is performed along the real-axis, and the subtraction of the imaginary parts is carried out along the j-axis. Now let $(4 + j5) - (1 + j4) = 3 + j$ be represented by the phasor $\overrightarrow{OQ}$.

Fig. **9.8**

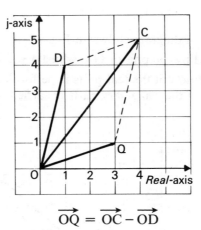

It follows that
$$\overrightarrow{OQ} = \overrightarrow{OC} - \overrightarrow{OD}$$

As for phasor addition, the subtraction of phasors is similar to the subtraction of vectors.

The Polar Form of a Complex Number

Let z denote the complex number represented by the phasor $\overrightarrow{OP}$ shown in Fig. 9.9. Then from the right-angled triangle PMO we have

$$z = x + jy$$
$$= r \cos \theta + j(r \sin \theta)$$
$$= r(\cos \theta + j \sin \theta)$$

Fig. **9.9**

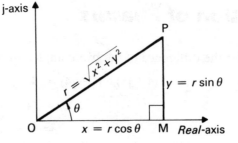

The expression $r(\cos \theta + j \sin \theta)$ is known as the **polar form** of the complex number z. Using conventional notation it may be shown abbreviated as $r\underline{/\theta}$.

r is called the **modulus** of the complex number z and is denoted by mod z or $|z|$.

Hence, from the diagram, $|z| = r = \sqrt{x^2 + y^2}$, using the theorem of Pythagoras for right-angled triangle PMO.

It should be noted that the plural of 'modulus' is 'moduli'.

The angle θ is called the **argument** (or amplitude) of the complex number z, and is denoted by arg z (or amp z).

Hence $\qquad\qquad\qquad$ arg $z = \theta$

and, from the diagram $\qquad\qquad \tan \theta = \dfrac{y}{x}$

There are an infinite number of angles whose tangents are the same, and so it is necessary to define which value of θ to state when solving the equation $\tan \theta = \dfrac{y}{x}$. It is called the **principal value** of the angle and lies between $+180°$ and $-180°$.

We recommend that, when finding the polar form of a complex number, you should sketch it on an Argand diagram. This will help you to avoid a common error of giving an incorrect value of the angle.

Example 9.5

Find the modulus and argument of the complex number $3 + j4$ and express the complex number in polar form.

Let $z = 3 + j4$ which is shown in the Argand diagram in Fig. 9.10.

 Fig. **9.10**

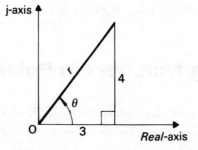

Then $\qquad |z| = r = \sqrt{3^2 + 4^2} = 5$

and $\qquad \tan \theta = \frac{4}{3} = 1.3333$

$\therefore \qquad\qquad \theta = 53.13°$

Hence in polar form

$$z = 5(\cos 53.13° + j \sin 53.13°)$$

or $\qquad\qquad z = 5\underline{/53.13°}$

Example 9.6

Show the complex number $z = 5\underline{/-150°}$ on an Argand diagram, and find z in algebraic form.

Now z is represented by phasor $\overrightarrow{OP}$ in Fig. 9.11. It should be noted that since the angle is negative it is measured in a clockwise direction from the real-axis datum.

Fig. **9.11**

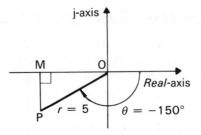

In order to express z in algebraic form we need to find the lengths MO and MP. We use the right-angled triangle PMO in which $\widehat{POM} = 180° - 150° = 30°$.

Now $\qquad MO = PO \cos \widehat{POM} = 5 \cos 30° = 4.33$

and $\qquad MP = PO \sin \widehat{POM} = 5 \sin 30° = 2.50$

Hence, in algebraic form, the complex number $z = -4.33 - j2.50$.

Multiplying Numbers in Polar Form

Consider the complex number $z_1 = r_1\underline{/\theta_1} = r_1(\cos \theta_1 + j \sin \theta_1)$
and another complex number $z_2 = r_2\underline{/\theta_2} = r_2(\cos \theta_2 + j \sin \theta_2)$

Then the product of these two complex numbers is

$$z_1 \times z_2 = r_1(\cos\theta_1 + j\sin\theta_1) \times r_2(\cos\theta_2 + j\sin\theta_2)$$

$$= r_1 r_2(\cos\theta_1 + j\sin\theta_1)(\cos\theta_2 + j\sin\theta_2)$$

$$= r_1 r_2(\cos\theta_1 \cos\theta_2 + j\sin\theta_1 \cos\theta_2 \\ + j\cos\theta_1 \sin\theta_2 + j^2 \sin\theta_1 \sin\theta_2)$$

$$= r_1 r_2[(\cos\theta_1 \cos\theta_2 - \sin\theta_1 \sin\theta_2) \\ + j(\sin\theta_1 \cos\theta_2 + \cos\theta_1 \sin\theta_2)]$$

$$= r_1 r_2\{\cos(\theta_1 + \theta_2) + j\sin(\theta_1 + \theta_2)\}^{\dagger}$$

$$= r_1 r_2 \underline{/\theta_1 + \theta_2}$$

> Hence to multiply two complex numbers we multiply their moduli and add their arguments.

For example, $6\underline{/17°} \times 3\underline{/35°} = 6 \times 3\underline{/17° + 35°}$

$$= 18\underline{/52°}$$

Square Root of a Complex Number

Let $z = r\underline{/\theta}$

$$= (\sqrt{r})(\sqrt{r})\,\underline{\left/\dfrac{\theta}{2} + \dfrac{\theta}{2}\right.}$$

and using the fact that $r_1 r_2 \underline{/\theta_1 + \theta_2} = (r_1\underline{/\theta_1}) \times (r_2\underline{/\theta_2})$

we have $z = \left(\sqrt{r}\,\underline{\left/\dfrac{\theta}{2}\right.}\right) \times \left(\sqrt{r}\,\underline{\left/\dfrac{\theta}{2}\right.}\right)$

$\therefore$ $z = \left(\sqrt{r}\,\underline{\left/\dfrac{\theta}{2}\right.}\right)^2$

$\therefore$ $\sqrt{z} = \sqrt{r}\,\underline{\left/\dfrac{\theta}{2}\right.}$

Thus

> The square root of a complex number is another complex number having a modulus equal to the square root of the given modulus, and argument half that of the given argument.

†See p. 95 for full explanation of the compound angle formulae.

Thus the square root of $36\underline{/24°}$ is $\sqrt{36}\left\lfloor\dfrac{24°}{2}\right.$ or $6\underline{/12°}$.

This is not the only square root but further discussion of these is beyond the scope of our studies at this stage.

If the square root of a complex number in algebraic form is required the number should first be put in polar form and the square root found as above. If needed, the square root may then be expressed in algebraic form.

Dividing Numbers in Polar Form

It can be shown that the division of two complex numbers, using a method similar to that for finding the product of two complex numbers, is given by

$$\frac{z_1}{z_2} = \frac{r_1\underline{/\theta_1}}{r_2\underline{/\theta_2}} = \frac{r_1}{r_2}\underline{/\theta_1-\theta_2}$$

> Hence if we divide two complex numbers we divide their moduli and subtract their arguments.

For example,

$$\frac{5\underline{/33.92°}}{3\underline{/-23.67°}} = \frac{5}{3}\underline{/(33.92°)-(-23.67°)}$$

$$= 1.67\underline{/33.92° + 23.67°}$$

$$= 1.67\underline{/57.59°}$$

Example 9.7

A simple circuit which has a resistance R in series with an inductive reactance X_L has an impedance Z given by the complex number

$$Z = R + jX_L$$

A simple circuit which has a resistance R in series with a capacitive reactance X_C has an impedance Z given by the complex number

$$Z = R - jX_C$$

Using the above relationships, find the resistance and the inductive or capacitive reactance for each of the following impedances:

(a) $8 + j12$

(b) $20 - j80$

(c) $40\underline{/25°}$

(d) $100\underline{/-20°}$

(a) Here $Z = 8 + j12$, and since it is of the form $Z = R + jX_L$ we can say that the resistance $R = 8$ and the inductive reactance $X_L = 12$.

(b) Here $Z = 20 - j80$, and since it is of the form $Z = R - jX_C$ we can say that the resistance $R = 20$ and the capacitive reactance $X_C = 80$.

(c) The complex number $Z = 40\underline{/25°}$ is shown on the Argand diagram in Fig. 9.12. If we let $Z = x + jy$, then from the diagram

$$x = 40 \cos 25°$$

$$= 36.3$$

and $$y = 40 \sin 25°$$

$$= 16.9$$

Hence $Z = 36.3 + j16.9$ which is of the form $Z = R + jX_L$ and we can say that resistance $R = 36.3$ and the inductive reactance $X_L = 16.9$.

Fig. **9.12**

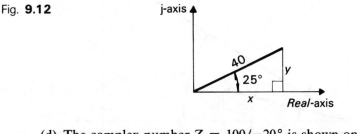

(d) The complex number $Z = 100\underline{/-20°}$ is shown on the Argand diagram in Fig. 9.13. If we let $Z = x + jy$, then from the diagram

$$x = 100 \cos 20°$$

$$= 94.0$$

and $$y = 100 \sin 20°$$

$$= 34.2$$

but we can see from the diagram that the y value is negative. Hence $Z = 94.0 - j34.2$, which is of the form $Z = R - jX_C$ and we can say that the resistance $R = 94.0$ and the capacitive reactance $X_C = 34.2$.

Fig. **9.13**

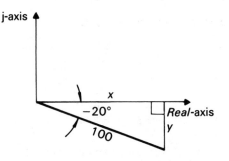

Example 9.8

The potential difference across a circuit is given by the complex number $V = 50 + j30$ volts, and the current is given by the complex number $I = 9 + j4$ amperes. Find:

(a) the phase difference (i.e. the angle ϕ in Fig. 9.14) between the phasors for V and I

(b) the power, given that power $= |V| \times |I| \times \cos \phi$ watts

Fig. **9.14**

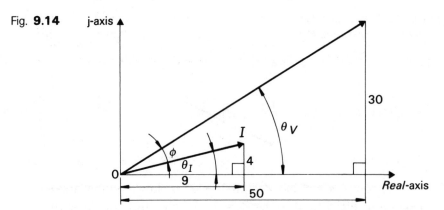

Fig. 9.14 shows a sketch of the Argand diagram showing the phasors for I and V. Phasors in electrical work are usually shown with arrows.

To find $V = 50 + j30$ in polar form

$$|V| = \sqrt{50^2 + 30^2}$$

$$= 58.3$$

and $$\tan \theta_V = \frac{30}{50}$$

$\therefore$ $$\theta_V = 30.96°$$

To find $I = 9 + j4$ in polar form

$$|I| = \sqrt{9^2 + 4^2}$$

$$= 9.8$$

and $$\tan \theta_I = \frac{4}{9}$$

$\therefore$ $$\theta_I = 23.96°$$

(a) The phase difference

$$\phi = \theta_V - \theta_I$$

$$= 30.96° - 23.96°$$

$$= 7°$$

(b) $$\text{Power} = |V| \times |I| \times \cos \phi$$

$$= 58.3 \times 9.8 \times \cos 7°$$

$$= 567 \text{ watts}$$

EXERCISE

1 Show, indicating each one clearly, the following complex numbers on a single Argand diagram: $4 + j3$, $-2 + j$, $3 - j4$, $-3.5 - j2$, $j3$ and $-j4$.

2 Find the moduli and arguments of the complex numbers $3 + j4$ and $4 - j3$.

3 If the complex number $z_1 = -3 + j2$ find $|z_1|$ and $\arg z_1$.

4 If the complex number $z_2 = -4 - j2$ find $|z_2|$ and $\arg z_2$.

5 Express each of the following complex numbers in polar form:

(a) $4 + j3$ (b) $3 - j4$ (c) $-3 + j3$
(d) $-2 - j$ (e) $j4$ (f) $-j3.5$

6 Convert the following complex numbers, which are given in polar form, into Cartesian form:

(a) $3\underline{/45°}$ (b) $5\underline{/154°}$ (c) $4.6\underline{/-20°}$
(d) $3.2\underline{/-120°}$

7 Simplify the following products of two complex numbers, given in polar form, expressing the answer in polar form:

(a) $8/30° \times 7/40°$

(b) $2/-20° \times 5/-30°$

(c) $5/120° \times 3/-30°$

(d) $7/-50° \times 3/-40°$

8 Simplify the following divisions of two complex numbers, given in polar form, expressing the answer in polar form:

(a) $\dfrac{8/20°}{3/50°}$

(b) $\dfrac{10/-40°}{5/20°}$

(c) $\dfrac{3/-15°}{5/-6°}$

(d) $\dfrac{1.7/35.28°}{0.6/-9.37°}$

9 Three complex numbers z_1, z_2 and z_3 are given in polar form by $z_1 = 3/35°$, $z_2 = 5/28°$ and $z_3 = 2/-50°$. Simplify:

(a) $z_1 \times z_2 \times z_3$, giving the answer in polar form.

(b) $\dfrac{z_1 \times z_2}{z_3}$, giving the answer in algebraic form.

10 If the complex number $z = 2 - j3$ express in polar form:

(a) $\dfrac{1}{z}$

(b) z^2

11 Find the square root of:

(a) $8/38°$ in polar form

(b) $2 + j3$ in algebraic form

12 The admittance Y of a circuit is given by $Y = \dfrac{1}{Z}$.

(a) If $Z = 3 + j5$ find Y in polar form.

(b) If $Z = 17.4/42°$ find Y in algebraic form.

13 Using the notation and information given in the data for the worked Example 9.7 of the text (p. 124), find the resistance and the inductive or capacitive reactance for each of the following impedances:

(a) $4.5 + 2.2j$

(b) $23 - 35j$

(c) $29.6/23.37°$

(d) $7/-12°$

14 The potential difference across a circuit is given by the complex number $V = 40 + j35$ volts and the current is given by the complex number $I = 6 + j3$ amperes. Sketch the appropriate phasors on an Argand diagram and find:

   (a) the phase difference (i.e. the angle ϕ) between the phasors for V and I
   (b) the power, given that power $= |V| \times |I| \times \cos \phi$

Differentiation

10

OBJECTIVES

1 Use the derivatives of the functions: ax^n, sin ax, cos ax, tan x, $\log_e x$, and e^{ax}.

2 Define the differential property of the exponential function.

3 Calculate the derivative at a point of the functions in **1**.

4 State the basic rules of differential calculus for the derivatives of sum, product, quotient, and function of a function.

5 Determine the derivatives of various combinations of any two of the functions in **1** using **4**.

6 Evaluate the derivatives in **5** at a given point.

Introduction

You may recall that the process of differentiation is a method of finding the rate of change of a function. The rate of change at a point on a curve may be found by determining the gradient of the tangent at that point. This may be achieved by either a theoretical or graphical method as follows.

Suppose we wish to find the gradient of the curve $y = x^2$ at the points where $x = 3$ and $x = -2$.

From a theoretical approach you may remember that if the function is of the form $y = x^n$, then the process of differentiating gives $\dfrac{dy}{dx} = nx^{n-1}$.

Hence if $\qquad\qquad y = x^2$

then $\qquad\qquad \dfrac{dy}{dx} = 2x$

and therefore when $x = 3$, the gradient of the tangent

$$\frac{dy}{dx} = 2(3) = +6$$

and when $x = -2$, the gradient of the tangent

$$\frac{dy}{dx} = 2(-2) = -4$$

Alternatively if we decide to use the graphical method we must first draw the graph as shown in Fig. 10.1.

Fig. **10.1**

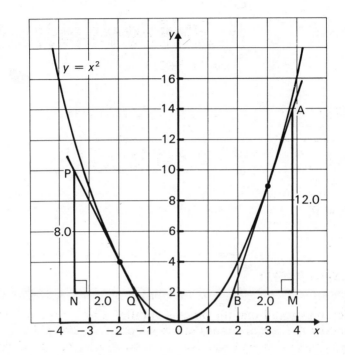

Tangents have been drawn at the given points, that is where $x = 3$ and $x = -2$, and then the gradients may be found by constructing suitable right-angled triangles.

Where $\quad x = 3 \quad$ the gradient $= \dfrac{AM}{BM} = \dfrac{12.0}{2.0} = +6$

and where $\quad x = -2 \quad$ the gradient $= \dfrac{PN}{QN} = -\dfrac{8.0}{2.0} = -4$

These results verify those obtained by the theoretical method.

Differentiation of a Sum

To differentiate an expression containing the sum of several terms, we differentiate each individual term separately.

Hence if $\qquad$ $y = x^4 + 2x^3 + 5x^2 + 7$

then $\qquad$ $\dfrac{dy}{dx} = 4x^3 + 6x^2 + 10x$

And if $\qquad$ $y = ax^3 + bx^2 - cx + d$

then $\qquad$ $\dfrac{dy}{dx} = 3ax^2 + 2bx - c$

And if $\qquad$ $y = \sqrt{x} + \dfrac{1}{\sqrt{x}} = x^{1/2} + x^{-1/2}$

then $\qquad$ $\dfrac{dy}{dx} = \tfrac{1}{2}x^{-1/2} + (-\tfrac{1}{2})x^{-3/2} = \dfrac{1}{2\sqrt{x}} - \dfrac{1}{2\sqrt{x^3}}$

And if $\qquad$ $y = 3.1x^{1.4} - \dfrac{3}{x} + 5 = 3.1x^{1.4} - 3x^{-1} + 5$

then $\qquad$ $\dfrac{dy}{dx} = (3.1)(1.4)x^{0.4} - 3(-1)x^{-2} = 4.34x^{0.4} + \dfrac{3}{x^2}$

And if $\qquad$ $y = \dfrac{t^3 + t}{t^2} = \dfrac{t^3}{t^2} + \dfrac{t}{t^2} = t + t^{-1}$

then $\qquad$ $\dfrac{dy}{dt} = 1 + (-1)t^{-2} = 1 - \dfrac{1}{t^2}$

EXERCISE

10·1

1 Find $\dfrac{dy}{dx}$ if $y = 5x^3 + 7x^2 - x - 1$.

2 If $s = 7\sqrt{t} - 6t^{0.3}$, find an expression for $\dfrac{ds}{dt}$.

3 Find by a theoretical method the value of $\dfrac{dy}{dx}$ when $x = 2$
for the curve $y = x - \dfrac{1}{x}$.

4 Find the expression for $\dfrac{dy}{du}$ if $y = \dfrac{u + u^2}{u}$.

5 Find graphically the gradient of the curve $y = x^2 + x + 2$ at the points where $x = +2$ and $x = -2$. Check the result by differentiation.

Functions

If two variables x and y are connected so that the value of y depends upon the value allocated to x, then y is said to be a **function** of x. Thus if $y = 3x^2 + 4x - 7$, when $x = 5$, $y = 3 \times 5^2 + 4 \times 5 - 7 = 88$. Since the value of y depends upon the value allocated to x, y then is a function of x.

Differentiation Using 'Function of a Function'

Up to now we have only learnt how to differentiate comparatively simple expressions such as $y = 3x^{4.5}$. For more difficult expressions such as $y = \sqrt{(x^3 + 3x - 9)}$ we make a substitution.

If we put $u = x^3 + 3x - 9$ then $y = \sqrt{u}$.

Now y is a function of u, and since u is a function of x, it follows that y is a function of a function of x.

This all sounds rather complicated and the words 'function of a function' should be noted in case you meet them again, but we prefer to use the expression 'differentiation by substitution'.

Differentiation by Substitution

A substitution method is often used for differentiating the more complicated expressions, together with the formula

$$\boxed{\frac{dy}{dx} = \frac{dy}{du} \times \frac{du}{dx}}$$

Example 10.1

Find $\dfrac{dy}{dx}$ if $y = (x^2 - x)^9$.

We have $y = (x^2 - x)^9$

Then $y = u^9$ where $u = x^2 - x$

$\therefore$ $\dfrac{dy}{du} = 9u^8$ and $\dfrac{du}{dx} = 2x - 1$

But $\dfrac{dy}{dx} = \dfrac{dy}{du} \times \dfrac{du}{dx}$

$\therefore$ $\dfrac{dy}{dx} = 9u^8 \times (2x - 1)$

The differentiation has now been completed and it only remains to put u in terms of x by using our original substitution $u = x^2 - x$.

Hence $\dfrac{dy}{dx} = 9(x^2 - x)^8 (2x - 1)$

Example 10.2

Find $\dfrac{d}{dx}\left(\sqrt{(1 - 5x^3)} \right)$.

$\dfrac{d}{dx}\left(\sqrt{(1 - 5x^3)} \right)$ is called the differential coefficient of $\sqrt{(1 - 5x^3)}$ with respect to x. This simply means that we have to differentiate the expression with respect to x. If we let $y = \sqrt{(1 - 5x^3)}$, then the problem is to find $\dfrac{dy}{dx}$.

Let $y = \sqrt{(1 - 5x^3)}$

i.e. $y = (1 - 5x^3)^{1/2}$

Then $y = u^{1/2}$ where $u = 1 - 5x^3$

$\therefore$ $\dfrac{dy}{du} = \tfrac{1}{2}u^{-1/2}$ and $\dfrac{du}{dx} = -15x^2$

But $\dfrac{dy}{dx} = \dfrac{dy}{du} \times \dfrac{du}{dx}$

$\therefore \qquad \dfrac{dy}{dx} = \tfrac{1}{2}u^{-1/2} \times (-15x^2)$

$\therefore \qquad \dfrac{dy}{dx} = \tfrac{1}{2}(1 - 5x^3)^{-1/2}(-15x^2)$

$\qquad\qquad = -\dfrac{15}{2}x^2(1 - 5x^3)^{-1/2}$

$\qquad\qquad = -\dfrac{15x^2}{2\sqrt{1 - 5x^3})}$

Hence $\qquad \dfrac{d}{dx}\left(\sqrt{(1 - 5x^3)}\right) = -\dfrac{15x^2}{2\sqrt{(1 - 5x^3)}}$

Differentiation of a Function of a Function by Recognition

Consider $y = (\quad)^n$ where any function of x can be written inside the bracket. Then differentiating with respect to x, we have

$$\frac{dy}{dx} = \frac{dy}{d(\quad)} \times \frac{d(\quad)}{dx}$$

Thus to differentiate an expression of the type $(\quad)^n$, first differentiate the bracket, treating it as a term similar to x^n. Then differentiate the function x inside the bracket. Finally, to obtain an expression for $\dfrac{dy}{dx}$, multiply these two results together.

Example 10.3

Find $\dfrac{dy}{dx}$ if $y = (x^2 - 5x + 3)^5$.

Differentiating the bracket as a whole we have

$$\frac{dy}{d(\quad)} = 5(x^2 - 5x + 3)^4 \qquad\qquad [1]$$

Also the function inside the bracket is $x^2 - 5x + 3$.

Differentiating this gives

$$\frac{d(\ )}{dx} = 2x - 5 \qquad [2]$$

Thus multiplying the results [1] and [2] together gives

$$\frac{dy}{dx} = 5(x^2 - 5x + 3)^4 \times (2x - 5)$$

$$= 5(2x - 5)(x^2 - 5x + 3)^4$$

Hence by recognising the method we can differentiate directly. If, for example, $y = (x^2 - 3x)^7$

then $$\frac{dy}{dx} = 7(x^2 - 3x)^6 \times (2x - 3)$$

$$= 7(2x - 3)(x^2 - 3x)^6$$

To Find the Rate of Change of sin θ, i.e. $\frac{d}{d\theta}$ (sin θ)

The rate of change of a curve at any point is the gradient of the tangent at that point. We shall, therefore, find the gradient at various points on the graph of sin θ and then plot the values of these gradients to obtain a new graph.

It is suggested that the reader follows the method given, plotting his or her own curves on graph paper.

First, we plot the graph of $y = \sin \theta$ from $\theta = 0°$ to $\theta = 90°$ using values of sin θ which may be obtained from your calculator. The curve is shown in Fig. 10.2.

Consider point P on the curve, where $\theta = 45°$, and draw the tangent APM.

We can find the gradient of the tangent by constructing a suitable right-angled triangle AMN (which should be as large as conveniently possible for accuracy) and finding the value of $\frac{MN}{AN}$.

Fig. **10.2**

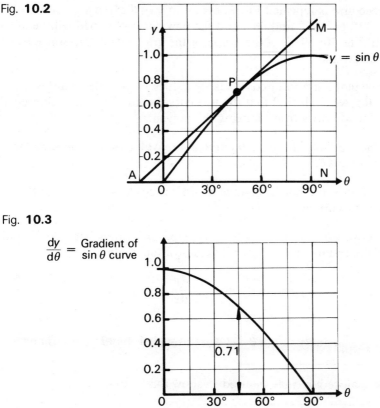

Fig. **10.3**

Using the scale on the y-axis gives MN = 1.29 by measurement, and using the scale on the θ-axis gives AN = 104° by measurement.

In calculations of this type it is necessary to obtain AN in radians. Remembering that

$$360° = 2\pi \text{ radians}$$

gives $$1° = \frac{2\pi}{360} \text{ radians}$$

∴ $$104° = \frac{2\pi}{360} \times 104 = 1.81 \text{ radians}$$

Hence   Gradient at P = $\dfrac{MN}{AN} = \dfrac{1.29}{1.81} = 0.71$

The value 0.71 is used as the y-value at θ = 45° to plot a point on a new graph using the same scales as before. This new graph could be plotted on the same axes as y = sin θ but for clarity it has been shown on new axes in Fig. 10.3.

This procedure is repeated for points on the $\sin \theta$ curve at θ values of $0°$, $15°$, $30°$, $60°$, $75°$ and $90°$, and the new curve obtained will be as shown in Fig. 10.3. This is the graph of the gradients of the sine curve at various points.

If we now plot a graph of $\cos \theta$, taking values from tables, on the axes in Fig. 10.3, we shall find that the two curves coincide — any difference will be due to errors from drawing the tangents.

Hence the gradient of the $\sin \theta$ curve at any value of θ is the same as the value of $\cos \theta$.

In other words, the rate of change of $\sin \theta$ is $\cos \theta$, provided that the angle θ is in radians.

In the above work we have only considered the graphs between $0°$ and $90°$ but the results are true for all values of the angle.

Hence if $\qquad y = \sin \theta \quad$ then $\quad \dfrac{dy}{dx} = \cos \theta$

or $\qquad \boxed{\dfrac{d}{d\theta}(\sin \theta) = \cos \theta} \qquad$ provided that θ is in radians.

The same procedure may be used to show that

$$\boxed{\dfrac{d}{d\theta}(\cos \theta) = -\sin \theta} \qquad \text{provided that } \theta \text{ is in radians.}$$

Example 10.4

Find $\dfrac{d}{d\theta}(\sin 7\theta)$.

Let $\qquad\qquad\qquad y = \sin 7\theta$

Then $\qquad\qquad\quad y = \sin u \qquad\quad$ where $\quad u = 7\theta$

$\therefore \qquad\qquad\qquad \dfrac{dy}{du} = \cos u \qquad\quad$ and $\quad \dfrac{du}{d\theta} = 7$

But $\qquad\qquad\quad \dfrac{dy}{d\theta} = \dfrac{dy}{du} \times \dfrac{du}{d\theta}$

$$\therefore \qquad \frac{dy}{d\theta} = (\cos u) \times 7 = 7 \cos u = 7 \cos 7\theta$$

$$\therefore \qquad \frac{d}{d\theta}(\sin 7\theta) = 7 \cos 7\theta$$

Example 10.5

Find $\dfrac{d}{d\theta}(\cos 4\theta)$.

Let $\qquad\qquad\qquad y = \cos 4\theta$

Then $\qquad\qquad\qquad y = \cos u \qquad$ where $\quad u = 4\theta$

$$\therefore \qquad\qquad \frac{dy}{du} = -\sin u \qquad \text{and} \quad \frac{du}{d\theta} = 4$$

But $\qquad\qquad\qquad \dfrac{dy}{d\theta} = \dfrac{dy}{du} \times \dfrac{du}{d\theta}$

$$\therefore \qquad\qquad \frac{dy}{d\theta} = (-\sin u) \times 4 = -4 \sin u = -4 \sin 4\theta$$

$$\therefore \qquad \frac{d}{d\theta}(\cos 4\theta) = -4 \sin 4\theta$$

In general

$$\boxed{\begin{aligned} \frac{d}{dx}(\sin ax) &= a \cos ax \\[2mm] \frac{d}{dx}(\cos ax) &= -a \sin ax \end{aligned}}$$

and

Example 10.6

Find $\dfrac{d}{dt}\left[\cos\left(2t - \dfrac{3\pi}{2}\right)\right]$.

Let $\qquad\qquad\qquad y = \cos\left(2t - \dfrac{3\pi}{2}\right)$

Then $\qquad\qquad\qquad y = \cos u \qquad$ where $\quad u = 2t - \dfrac{3\pi}{2}$

$\therefore$ $$\frac{dy}{du} = -\sin u \qquad \text{and} \qquad \frac{du}{dt} = 2$$

But $$\frac{dy}{dt} = \frac{dy}{du} \times \frac{du}{dt}$$

$\therefore$ $$\frac{dy}{dt} = (-\sin u) \times 2 = -2\sin u$$

$$= -2\sin\left(2t - \frac{3\pi}{2}\right)$$

$\therefore$ $$\frac{d}{dt}\left[\cos\left(2t - \frac{3\pi}{2}\right)\right] = -2\sin\left(2t - \frac{3\pi}{2}\right)$$

The Differential Coefficient of $\log_e x$, i.e. $\frac{d}{dx}(\log_e x)$

In higher mathematics all logarithms are taken to the base e, where e = 2.718 28. Logarithms to this base are often called natural logarithms. They are also called Napierian or hyperbolic logarithms and are given as $\log_e$ or ln.

Again the graphical method of differentiation may be used. Fig. 10.4 shows the graph of $\log_e x$.

The reader may find it instructive to plot the curve of $y = \log_e x$ as shown in Fig. 10.4 and follow the procedure, as used previously, of drawing tangents at various points. The values of their gradients are then plotted, and a curve will result as shown in Fig. 10.5.

If a graph of $\frac{1}{x}$ is plotted on the same axes as in Fig. 10.5, it will be found to coincide with the gradient curve, that is, the $\frac{dy}{dx}$ graph.

Hence $$\boxed{\frac{d}{dx}(\log_e x) = \frac{1}{x}}$$

Fig. **10.4**

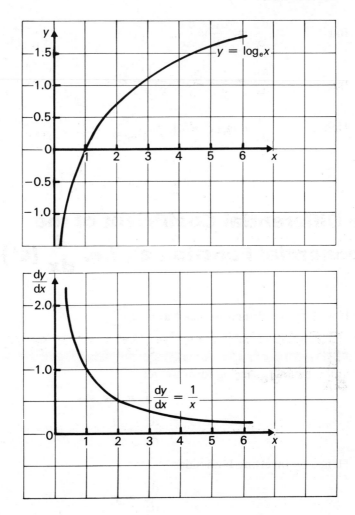

Fig. **10.5**

Example 10.7

Find $\dfrac{d}{dx}[\log_e(x^2+5)]$

Let $\qquad\qquad y = \log_e(x^2+5)$

Then $\qquad\qquad y = \log_e u \qquad$ where $\quad u = x^2+5$

$\therefore \qquad\qquad \dfrac{dy}{du} = \dfrac{1}{u} \qquad\qquad$ and $\quad \dfrac{du}{dx} = 2x$

But $\qquad \dfrac{dy}{dx} = \dfrac{dy}{du} \times \dfrac{du}{dx}$

$\therefore \qquad \dfrac{dy}{dx} = \dfrac{1}{u} \times 2x = \dfrac{1}{x^2 + 5} \times 2x$

Hence $\qquad \dfrac{d}{dx}[\log_e(x^2 + 5)] = \dfrac{2x}{x^2 + 5}$

The Differential Coefficient of the Exponential Function e^x, i.e. $\dfrac{d}{dx}(e^x)$

If we let $y = e^x$ then we need to find $\dfrac{dy}{dx}$.

Then $\qquad\qquad e^x = y$

and rearranging in log form we get

$$x = \log_e y$$

$\therefore \qquad\qquad \dfrac{dx}{dy} = \dfrac{1}{y}$

Hence by inverting both sides

$$\dfrac{dy}{dx} = y$$

But we know that $y = e^x$ and

$\therefore \qquad\qquad \dfrac{dy}{dx} = e^x$

Hence $\qquad\qquad \dfrac{d}{dx}(e^x) = e^x$

This result illustrates an important property of the exponential function, namely:

> The exponential function e^x has a differential coefficient of e^x which, therefore, is equal to the function itself.

Example 10.8

Find $\dfrac{d}{dx}(e^{6x})$.

Let	$y = e^{6x}$	
Then	$y = e^u$	where $u = 6x$

$\therefore \qquad\qquad \dfrac{dy}{du} = e^u \qquad\qquad$ and $\dfrac{du}{dx} = 6$

But $\qquad\qquad \dfrac{dy}{dx} = \dfrac{dy}{du} \times \dfrac{du}{dx}$

$\therefore \qquad\qquad \dfrac{dy}{dx} = e^u \times 6 = e^{6x} \times 6 = 6e^{6x}$

Hence $\qquad\qquad\qquad \dfrac{d}{dx}(e^{6x}) = 6e^{6x}$

Example 10.9

Find $\dfrac{d}{dx}\left(\dfrac{1}{e^{3x}}\right)$.

Let	$y = \dfrac{1}{e^{3x}}$	
or	$y = e^{-3x}$	
Then	$y = e^u$	where $u = -3x$

$\therefore \qquad\qquad \dfrac{dy}{du} = e^u \qquad\qquad$ and $\dfrac{du}{dx} = -3$

But $\qquad\qquad \dfrac{dy}{dx} = \dfrac{dy}{du} \times \dfrac{du}{dx}$

$\therefore \qquad\qquad \dfrac{dy}{dx} = e^u \times (-3) = -3e^u = -3e^{-3x}$

$\therefore \qquad \dfrac{d}{dx}\left(\dfrac{1}{e^{3x}}\right) = \dfrac{d}{dx}(e^{-3x}) = -3e^{-3x}$

In general $\qquad\qquad \boxed{\dfrac{d}{dx}(e^{ax}) = ae^{ax}}$

We may now summarise differential coefficients of the more common functions:

y	$\dfrac{dy}{dx}$
ax^n	anx^{n-1}
$\sin ax$	$a \cos ax$
$\cos ax$	$-a \sin ax$
$\log_e x$	$\dfrac{1}{x}$
e^{ax}	$a\,e^{ax}$

At this stage we may use the results in the table to differentiate many functions by 'recognition' rather than using the method of substitution.

For example, $\dfrac{d}{d\theta}(\sin 3\theta) = 3 \cos 3\theta$

or $\dfrac{d}{dx}(e^{-3x}) = -3e^{-3x}$

However, if the functions to be differentiated are not similar to those shown in the table, perhaps because they are more complicated, the method of substitution should be used.

Example 10.10 which follows shows how both the methods of substitution and recognition may be used together to differentiate a more complicated function.

Example 10.10

Find $\dfrac{dy}{dx}$ if $y = \sin^3 5x$.

We have $y = (\sin 5x)^3$

Then $y = u^3$ where $u = \sin 5x$

$\therefore$ $\dfrac{dy}{du} = 3u^2$ and $\dfrac{du}{dx} = 5 \cos 5x$

But $\qquad \dfrac{dy}{dx} = \dfrac{dy}{du} \times \dfrac{du}{dx}$

$\therefore \qquad \dfrac{dy}{dx} = 3u^2 \times 5\cos 5x = 3\sin^2 5x \times 5\cos 5x$

$\therefore \qquad \dfrac{dy}{dx} = 15\sin^2 5x \cos 5x$

EXERCISE

10·2

Differentiate with respect to x:

1 $(3x + 1)^2$ 　　　　　**2** $(2 - 5x)^3$ 　　　　　**3** $(1 - 4x)^{1/2}$

4 $(2 - 5x)^{3/2}$ 　　　**5** $\dfrac{1}{4x^2 + 3}$ 　　　**6** $\sin(3x + 4)$

7 $\cos(2 - 5x)$ 　　　**8** $\sin^2 4x$ 　　　**9** $\dfrac{1}{\cos^3 7x}$

10 $\sin\left(2x + \dfrac{\pi}{2}\right)$ 　　**11** $\cos^3 x$ 　　**12** $\dfrac{1}{\sin x}$

13 $\log_e 9x$ 　　**14** $9\log_e\left(\dfrac{5}{x}\right)$ 　　**15** $\tfrac{1}{4}\log_e(2x - 7)$

16 $\dfrac{1}{e^x}$ 　　**17** $2e^{3x + 4}$ 　　**18** $\dfrac{1}{e^{2 - 8x}}$

19 Find $\dfrac{d}{dt}\left(\dfrac{1}{\sqrt[3]{1 - 2t}}\right)$. 　　**20** Find $\dfrac{d}{d\theta}[\sin(\tfrac{3}{4}\theta - \pi)]$.

21 Find $\dfrac{d}{d\phi}\left(\dfrac{1}{\cos(\pi - \phi)}\right)$. 　　**22** Find $\dfrac{d}{dx}\left(\log_e\dfrac{1}{\sqrt{x}}\right)$.

23 Find $\dfrac{d}{dt}(Be^{kt - b})$. 　　**24** Find $\dfrac{d}{dx}(\sqrt[3]{e^{1 - x}})$.

Differentiation of a Product

If $y = u \times v$, where u and v are functions of x, we must use the formula

$$\boxed{\frac{dy}{dx} = v\frac{du}{dx} + u\frac{dv}{dx}}$$

Example 10.11

Find $\dfrac{d}{dx}(x^3 \sin 2x)$.

Let $\quad y = x^3 \sin 2x$

Then $\quad y = u \times v \quad$ where $\quad u = x^3 \quad$ and $\quad v = \sin 2x$

$$\therefore \quad \frac{du}{dx} = 3x^2 \text{ and } \frac{dv}{dx} = 2\cos 2x$$

But $\quad \dfrac{dy}{dx} = v\dfrac{du}{dx} + u\dfrac{dv}{dx}$

$$\therefore \quad \frac{dy}{dx} = (\sin 2x)3x^2 + x^3(2\cos 2x) = x^2(3\sin 2x + 2x\cos 2x)$$

Then $\qquad \dfrac{d}{dx}(x^3 \sin 2x) = x^2(3\sin 2x + 2x\cos 2x)$

Example 10.12

Differentiate $(x^2 + 1)\log_e x$ with respect to x.

If we let $\quad y = (x^2 + 1)\log_e x \quad$ then the problem is to find $\dfrac{dy}{dx}$.

Then $\quad y = u \times v \quad$ where $\quad u = x^2 + 1 \quad$ and $\quad v = \log_e x$

$$\therefore \quad \frac{du}{dx} = 2x \qquad \text{and } \frac{dv}{dx} = \frac{1}{x}$$

But $\quad \dfrac{dy}{dx} = v\dfrac{du}{dx} + u\dfrac{dv}{dx}$

$$\therefore \quad \frac{dy}{dx} = (\log_e x)2x + (x^2 + 1)\frac{1}{x} = 2x(\log_e x) + x + \frac{1}{x}$$

Differentiation of a Quotient

If $y = \dfrac{u}{v}$, where u and v are functions of x, we must use the formula

$$\frac{dy}{dx} = \frac{v\dfrac{du}{dx} - u\dfrac{dv}{dx}}{v^2}$$

Example 10.13

Find $\dfrac{dy}{dx}$ if $y = \dfrac{e^{2x}}{x+3}$.

We have $y = \dfrac{e^{2x}}{x+3}$

Let $y = \dfrac{u}{v}$ where $u = e^{2x}$ and $v = x+3$

$$\therefore \qquad \frac{du}{dx} = 2e^{2x} \quad \text{and} \quad \frac{dv}{dx} = 1$$

But $\dfrac{dy}{dx} = \dfrac{v\dfrac{du}{dx} - u\dfrac{dv}{dx}}{v^2}$

$\therefore \qquad \dfrac{dy}{dx} = \dfrac{(x+3)2e^{2x} - e^{2x} \times 1}{(x+3)^2}$

$\qquad\qquad = \dfrac{e^{2x}(2x+6-1)}{(x+3)^2}$

$\qquad\qquad = \dfrac{(2x+5)e^{2x}}{(x+3)^2}$

Example 10.14

Find $\dfrac{d}{d\theta}(\tan\theta)$.

Let $y = \tan\theta$

or $y = \dfrac{\sin\theta}{\cos\theta}$

Then $y = \dfrac{u}{v}$ where $u = \sin \theta$ and $v = \cos \theta$

$$\therefore \qquad \frac{du}{d\theta} = \cos \theta \quad \text{and} \quad \frac{dv}{d\theta} = -\sin \theta$$

But $\dfrac{dy}{d\theta} = \dfrac{v\dfrac{du}{d\theta} - u\dfrac{dv}{d\theta}}{v^2}$

$\therefore \qquad \dfrac{dy}{d\theta} = \dfrac{(\cos \theta)(\cos \theta) - (\sin \theta)(-\sin \theta)}{\cos^2 \theta}$

$\qquad\qquad = \dfrac{\cos^2 \theta + \sin^2 \theta}{\cos^2 \theta}$

Using the identity $\sin^2 \theta + \cos^2 \theta = 1$,

then $\dfrac{dy}{d\theta} = \dfrac{1}{\cos^2 \theta} = \sec^2 \theta$

Hence $\boxed{\dfrac{d}{d\theta}(\tan \theta) = \sec^2 \theta}$ provided that θ is in radians.

EXERCISE 10·3

1 Differentiate with respect to x:

   (a) $x \sin x$ (b) $e^x \tan x$ (c) $x \log_e x$

2 Find $\dfrac{d}{dt}(\sin t \cos t)$.   3 Find $\dfrac{d}{d\theta}(\sin 2\theta \tan \theta)$.

4 Find $\dfrac{d}{dm}(e^{4m} \cos 3m)$.   5 Find $\dfrac{d}{dx}(3x^2 \log_e x)$.

6 Find $\dfrac{d}{dt}[6e^{3t}(t^2 - 1)]$ 7 $\dfrac{d}{dz}[(z - 3z^2) \log_e z]$.

8 Differentiate with respect to x:

   (a) $\dfrac{x}{1-x}$ (b) $\dfrac{\log_e x}{x^2}$ (c) $\dfrac{e^x}{\sin 2x}$

9 Find $\dfrac{d}{dz}\left(\dfrac{z+2}{3-4z}\right)$.

10 Find $\dfrac{d}{dt}\left(\dfrac{\cos 2t}{e^{2t}}\right)$.

11 Find $\dfrac{d}{d\theta}(\cos\theta)$. $\left(Hint:\ \text{Use the identity } \cot\theta = \dfrac{\cos\theta}{\sin\theta}.\right)$

Numerical Values of Differential Coefficients

Example 10.15

Find the value of $\dfrac{dy}{dx}$ for the curve $y = \dfrac{1}{\sqrt{x}} - 3\log_e x$ at the point where $x = 2.3$.

We have

$$y = x^{-\frac{1}{2}} - 3\log_e x$$

$$\therefore \qquad \frac{dy}{dx} = -\tfrac{1}{2}x^{-3/2} - \frac{3}{x}$$

It is often difficult to decide how much simplification of an expression will help in finding its numerical value when a particular value of x is substituted. In this case the expression may be rewritten as:

$$\frac{dy}{dx} = -\frac{1}{2(\sqrt{x})^3} - \frac{3}{x}$$

Hence when $\qquad x = 2.3$

then $\qquad \dfrac{dy}{dx} = -\dfrac{1}{2(\sqrt{2.3})^3} - \dfrac{3}{2.3} = -1.45$

It may well be argued that if a scientific calculator is available the value of x may just as well be substituted into the original expression for $\dfrac{dy}{dx}$ without any simplification.

This would give $\dfrac{dy}{dx} = \tfrac{1}{2}(2.3)^{-1.5} - \dfrac{3}{2.3}$ which can be evaluated as easily as the 'simplified' arrangement.

It may, however, be more difficult to detect a computation error when making a rough check of the answer, which is why expressions with positive indices are often preferred.

Example 10.16

A curve is given in the form $y = 3 \sin 2\theta - 5 \tan \theta$ when θ is in radians. Find the gradient of the curve at the point where θ has a value equivalent to $34°$.

The gradient of the curve is given by $\dfrac{dy}{d\theta}$.

We have
$$y = 3 \sin 2\theta - 5 \tan \theta$$

$$\frac{dy}{d\theta} = 3 \times 2 \cos 2\theta - 5 \sec^2 \theta$$

Substituting the value $\theta = 34°$

gives
$$\frac{dy}{d\theta} = 6 \cos (2 \times 34)° - 5 \sec^2 (34)°$$

$$= 6 \cos 68° - 5(\sec 34°)^2$$

$$= -5.03$$

Example 10.17

If $y = \frac{1}{2}(e^{3t} + e^{-3t})$ find the value of $\dfrac{dy}{dt}$ when $t = -0.63$.

We have
$$y = \frac{1}{2}(e^{3t} + e^{-3t})$$

$$\frac{dy}{dt} = \frac{1}{2}[3e^{3t} + (-3)e^{-3t}]$$

Substituting the value $t = -0.63$

gives
$$\frac{dy}{dt} = \frac{3}{2}[e^{3(-0.63)} - e^{-3(-0.63)}]$$

$$= \frac{3}{2}[e^{-1.89} - e^{1.89}]$$

$$= \frac{3}{2}[0.151 - 6.619]$$

$$= -9.70$$

Example 10.18

Find the gradient of the curve $\dfrac{\cos x}{x}$ at the point where $x = 0.25$.

The gradient of the curve is given by $\dfrac{dy}{dx}$ if we let

$$y = \frac{\cos x}{x}.$$

Then $\quad y = \dfrac{u}{v} \quad$ where $\quad u = \cos x \quad$ and $\quad v = x$

$$\therefore \quad \frac{du}{dx} = -\sin x \quad \text{and} \quad \frac{dv}{dx} = 1$$

But $\quad \dfrac{dy}{dx} = \dfrac{v\dfrac{du}{dx} - u\dfrac{dy}{dx}}{v^2}$

$\therefore \quad \dfrac{dy}{dx} = \dfrac{x(-\sin x) - (\cos x)1}{x^2}$

$$= \frac{-x \sin x - \cos x}{x^2}$$

If we now substitute the value $x = 0.25$

then $\quad \dfrac{dy}{dx} = \dfrac{-(0.25)(\sin 0.25) - (\cos 0.25)}{(0.25)^2}$

The value 0.25 must be treated as radians when substituted into trigonometrical functions such as $\sin x$ and $\cos x$.

If a scientific calculator is used it is usually possible to set it to accept radians and give trigonometrical ratios directly.

Then $\quad \dfrac{dy}{dx} = -16.5$

EXERCISE

10·4

1 If $y = 3x^2 - \dfrac{7}{x^2} + \sqrt{x}$, find the value of $\dfrac{dy}{dx}$ if $x = 3.5$.

2 If $y = 5 \sin 2\theta + 3 \cos \dfrac{\theta}{2}$, find the value of $\dfrac{dy}{d\theta}$ if
$\theta = 0.942$ radians.

3 Find the value of $\dfrac{dy}{dt}$ when $t = -0.1$ if $y = \frac{1}{2}(e^t - e^{-t})$.

4 If $x = 0.3$, find the value of $\dfrac{dy}{dx}$ when $y = \sqrt{(3 - 2x^2)}$.

5 If $y = \sin^4 2\theta$, find the value of $\dfrac{dy}{d\theta}$ when $\theta = \dfrac{3\pi}{2}$ radians.

6 A curve is given in the form $y = \tan(3\phi - \pi)$, where ϕ is
in radians. Find the gradient of the curve at the point
where ϕ has a value equivalent to $23.4°$.

7 If $y = 4 \log_e (1 - x)$, find the value of $\dfrac{dy}{dx}$ when $x = 0.32$.

8 Find the value of $\dfrac{dy}{dx}$ when $x = 2.9$ if $y = e^{(9 - 3x)}$.

9 Given that $y = (\sin x)(\cos x)$ find the value of $\dfrac{dy}{dx}$ when
$x = \dfrac{\pi}{6}$ radians.

10 If $y = \dfrac{1 + x^2}{x - 2}$, find the value of $\dfrac{dy}{dx}$ when $x = -1.25$.

Second Derivative

OBJECTIVES

1 State the notation for second derivatives as $\dfrac{d^2y}{dx^2}$ and similar forms, e.g. $\dfrac{d^2x}{dt^2}$.

2 Determine a second derivative by applying the basic rules of differential calculus to the simplified result of a first differentiation.

3 Evaluate a second derivative determined in **2** at a given point.

4 State that $\dfrac{ds}{dt}$ and $\dfrac{d^2s}{dt^2}$ express velocity and acceleration respectively.

5 Calculate the velocity and acceleration at a given time from an equation for displacement expressed in terms of time, using **4**.

Introduction

If

$$y = x^6$$

then

$$\frac{dy}{dx} = 6x^5$$

and if we differentiate this equation again with respect to x we obtain

$$\frac{d}{dx}\left(\frac{dy}{dx}\right) = \frac{d}{dx}(6x^5)$$

or

$$\frac{d^2y}{dx^2} = 30x^4$$

Now just as $\dfrac{dy}{dx}$ is called the **first differential coefficient**, or **first derivative**, of y with respect to x, so $\dfrac{d^2y}{dx^2}$ is called the **second differential coefficient**, or **second derivative**, of y with respect to x.

It should be noted that the figure 2 which occurs twice in $\dfrac{d^2y}{dx^2}$ is *not* an index but merely indicates that the original function has been differentiated twice. Hence $\dfrac{d^2y}{dx^2}$ is *not* the same as $\left(\dfrac{dy}{dx}\right)^2$.

153

Example 11.1

If $y = x^3 - 2x^2 + 3x - 7$ find $\dfrac{dy}{dx}$ and $\dfrac{d^2y}{dx^2}$.

Now
$$y = x^3 - 2x^2 + 3x - 7$$

$\therefore$
$$\frac{dy}{dx} = 3x^2 - 4x + 3$$

and
$$\frac{d^2y}{dx^2} = 6x - 4$$

Example 11.2

If $y = \sqrt{x} + \log_e x$ find the values of $\dfrac{dy}{dx}$ and $\dfrac{d^2y}{dx^2}$ when $x = 2$.

Now
$$y = x^{1/2} + \log_e x$$

$\therefore$
$$\frac{dy}{dx} = \tfrac{1}{2}x^{-1/2} + \frac{1}{x}$$

$$= \tfrac{1}{2}x^{-1/2} + x^{-1}$$

$\therefore$
$$\frac{d^2y}{dx^2} = -\tfrac{1}{4}x^{-3/2} + (-1)x^{-2} = -\tfrac{1}{4}x^{-3/2} - x^{-2}$$

Before substituting the numerical value of x we recommend that the expressions for $\dfrac{dy}{dx}$ and $\dfrac{d^2y}{dx^2}$ are transformed to give positive indices.

Although most electronic calculators will evaluate expressions with negative indices, mistakes may occur and it is more difficult to spot a computation error.

Now
$$\frac{dy}{dx} = \frac{1}{2x^{1/2}} + \frac{1}{x} = \frac{1}{2\sqrt{x}} + \frac{1}{x}$$

Hence when $x = 2$

$$\frac{dy}{dx} = \frac{1}{2\sqrt{2}} + \frac{1}{2} = 0.354 + 0.5 = 0.854$$

Also
$$\frac{d^2y}{dx^2} = -\frac{1}{4x^{3/2}} - \frac{1}{x^2} = -\frac{1}{4(\sqrt{x})^3} - \frac{1}{x^2}$$

and when $x = 2$

$$\frac{d^2y}{dx^2} = -\frac{1}{4(\sqrt{2})^3} - \frac{1}{2^2} = -0.088 - 0.250 = -0.338$$

Example 11.3

If $\theta = \dfrac{\pi}{2}$ find the values of $\dfrac{dy}{d\theta}$ and $\dfrac{d^2y}{d\theta^2}$ given that
$y = \sin 2\theta + \cos 3\theta$.

Now
$$y = \sin 2\theta + \cos 3\theta$$

$\therefore$
$$\frac{dy}{d\theta} = 2\cos 2\theta - 3\sin 3\theta$$

and
$$\frac{d^2y}{d\theta^2} = -4\sin 2\theta - 9\cos 3\theta$$

If a value of an angle is given in terms of π the units are radians.

Therefore, when $\theta = \dfrac{\pi}{2}$ we have

$$\frac{dy}{d\theta} = 2\cos 2\left(\frac{\pi}{2}\right) - 3\sin 3\left(\frac{\pi}{2}\right)$$

$$= 2\cos 180° - 3\sin 270° = 2(-1) - 3(-1) = 1$$

and when $\theta = \dfrac{\pi}{2}$ we have

$$\frac{d^2y}{d\theta^2} = -4\sin 2\left(\frac{\pi}{2}\right) - 9\cos 3\left(\frac{\pi}{2}\right) = -4(0) - 9(0) = 0$$

Example 11.4

If $y = \frac{1}{2}(e^{2t} + e^{-2t})$ find the values of $\dfrac{dy}{dt}$ and $\dfrac{d^2y}{dt^2}$ when $t = 0.61$.

We have
$$y = \tfrac{1}{2}e^{2t} + \tfrac{1}{2}e^{-2t}$$

$\therefore$
$$\frac{dy}{dt} = \tfrac{1}{2}(2e^{2t}) + \tfrac{1}{2}(-2e^{-2t})$$

$$= e^{2t} - e^{-2t}$$

Also $$\frac{d^2y}{dt^2} = 2e^{2t} - (-2)e^{-2t}$$

$$= 2e^{2t} + 2e^{-2t}$$

Hence when $t = 0.61$

$$\frac{dy}{dt} = e^{2(0.61)} - e^{-2(0.61)} = 3.09$$

and when $t = 0.61$

$$\frac{d^2y}{dt^2} = 2e^{2(0.61)} + 2e^{-2(0.61)} = 7.36$$

Example 11.5

If $y = \tan\theta$ find the value of $\dfrac{d^2y}{d\theta^2}$ when $\theta = 0.436$ radians.

We have $$y = \tan\theta$$

$\therefore$ $$\frac{dy}{d\theta} = \sec^2\theta$$

That is $$\frac{dy}{d\theta} = (\cos\theta)^{-2}$$

$\therefore$ $$\frac{d^2y}{d\theta^2} = 2(\sin\theta)(\cos\theta)^{-3}$$

You are left to check this second differentiation by differentiating $(\cos\theta)^{-2}$ by the method of substitution.

Now when $\theta = 0.436$ radians

$$\frac{d^2y}{d\theta^2} = \frac{2(\sin 0.436)}{(\cos 0.436)^3} = 1.135$$

EXERCISE

1 If $y = 3x^3 + 2x - 7$, find an expression for $\dfrac{d^2y}{dx^2}$ and also its value when $x = 3$.

2 Find the values of $\dfrac{dy}{dx}$ and $\dfrac{d^2y}{dx^2}$ when $x = -2$, given that $y = 5x^4 + 7x^2 + x$.

3 Given that $y = \dfrac{3t^5 + 2t}{t^2}$ find the value of $\dfrac{d^2y}{dt^2}$ when $t = 0.6$.

4 If $y = \log_e x$, find $\dfrac{d^2y}{dx^2}$ in terms of x, and also its value when $x = 1.9$.

5 If $z = 5\cos 3\theta$, find the value of $\dfrac{dz}{d\theta}$ and $\dfrac{d^2z}{d\theta^2}$ when $\theta = \dfrac{\pi}{2}$ radians.

6 Find the value of $\dfrac{d^2p}{d\phi^2}$, given that $p = 6\sin 4\phi$, where ϕ is in radians, when ϕ has a value equivalent to $28°$.

7 If $y = 2\cos\dfrac{\alpha}{4}$, find the value of $\dfrac{d^2y}{d\alpha^2}$ when $\alpha = 1.6$ radians.

8 When $t = 2$, find the value of $\dfrac{d^2v}{dt^2}$ given that $v = e^t + e^{-t}$.

9 If $y = -e^{4.6t}$, find the value of $\dfrac{dy}{dt}$ and $\dfrac{d^2y}{dt^2}$ when $t = 0$.

10 Find the value of $\dfrac{d^2u}{dm^2}$ if $u = \frac{1}{2}(e^{3m} - e^{-3m})$, given that $m = 1.3$.

11 If $y = x\log_e x$, find the value of $\dfrac{d^2y}{dx^2}$ if $x = 0.34$.

Velocity and Acceleration

Suppose that a vehicle starts from rest and travels 60 metres in 12 seconds. The average velocity may be found by dividing the total distance travelled by the total time taken, that is $\frac{60}{12} = 5$ m/s. This is *not* the *instantaneous* velocity, however, *at* a time of 12 seconds, but is the *average velocity* over the 12 seconds as calculated previously.

Fig. 11.1 shows a graph of distance s against time t. The average velocity over a period is given by the gradient of the chord which meets the curve at the extremes of the period. Thus in the diagram the gradient of the dotted chord QR gives the average velocity between $t = 2$ s and $t = 6$ s. It is found to be $\frac{13}{4} = 3.25$ m/s.

Fig. **11.1**

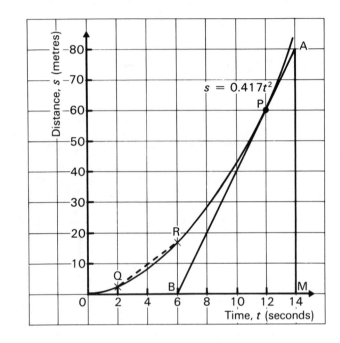

The velocity at any point is the rate of change of s with respect to t and may be found by finding the gradient of the curve at that point. In mathematical notation this is given by $\dfrac{ds}{dt}$.

Suppose we know that the relationship between s and t is

$$s = 0.417t^2$$

Then velocity $\qquad\qquad v = \dfrac{ds}{dt} = 0.834t$

and hence when $t = 12$ seconds, $v = 0.834 \times 12 = 10$ m/s.

This result may be found graphically by drawing the tangent to the curve of s against t at the point P and constructing a suitable right-angled triangle ABM.

Hence the velocity at P $= \dfrac{AM}{BM} = \dfrac{80}{8} = 10$ m/s which verifies the theoretical result.

Similarly, the rate of change of velocity with respect to time is called acceleration and is given by the gradient of the velocity–time graph at any point. In mathematical notation this is given by $\dfrac{dv}{dt}$.

Now
$$\frac{dv}{dt} = \frac{d}{dt}(v) = \frac{d}{dt}\left(\frac{ds}{dt}\right) = \frac{d^2s}{dt^2}$$

and so the acceleration a is given by either

$$\frac{dv}{dt} \quad \text{or} \quad \frac{d^2s}{dt^2}$$

The above reasoning was applied to linear motion, but it could also have been used for angular motion. The essential difference is that distance s is replaced by angle turned through, θ rad.

Both sets of results are summarised in Fig. 11.2.

Fig. **11.2**

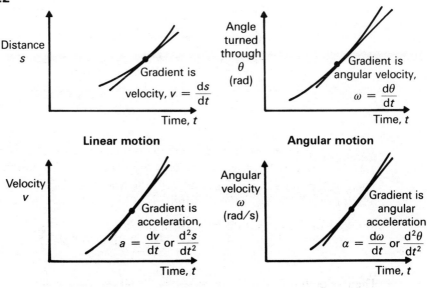

Example 11.6

A body moves a distance s metres in a time of t seconds so that $s = 2t^3 - 9t^2 + 12t + 6$. Find:

(a) its velocity after 3 seconds
(b) its acceleration after 3 seconds
(c) when the velocity is zero

We have $\qquad s = 2t^3 - 9t^2 + 12t + 6$

$\therefore \qquad\qquad \dfrac{ds}{dt} = 6t^2 - 18t + 12$

and $\qquad\qquad \dfrac{d^2s}{dt^2} = 12t - 18$

(a) When $t = 3$, then the velocity is

$$\dfrac{ds}{dt} = 6(3)^2 - 18(3) + 12 = 12 \text{ m/s}$$

(b) When $t = 3$, then the acceleration is $\dfrac{d^2s}{dt^2} = 12(3) - 18 = 18 \text{ m/s}^2$.

(c) When the velocity is zero then $\dfrac{ds}{dt} = 0$.

That is $\qquad\quad 6t^2 - 18t + 12 = 0$

$\therefore \qquad\qquad\quad t^2 - 3t + 2 = 0$

$\therefore \qquad\qquad (t-1)(t-2) = 0$

$\therefore$ either $\qquad\qquad t - 1 = 0 \quad$ or $\quad t - 2 = 0$

$\therefore$ either $\qquad\qquad t = 1$ second or $t = 2$ seconds

Example 11.7

The angle θ radians is connected with the time t seconds by the relationship $\theta = 20 + 5t^2 - t^3$. Find:

(a) the angular velocity when $t = 2$ seconds
(b) the value of t when the angular deceleration is 4 rad/s^2

We have $\qquad\qquad \theta = 20 + 5t^2 - t^3$

$\therefore \qquad\qquad \dfrac{d\theta}{dt} = 10t - 3t^2$

and $\qquad\qquad \dfrac{d^2\theta}{dt^2} = 10 - 6t$

(a) When $t = 2$, then the angular velocity

$$\dfrac{d\theta}{dt} = 10(2) - 3(2)^2 = 8 \text{ rad/s}$$

(b) An angular deceleration of 4 rad/s² may be called an angular acceleration of -4 rad/s².

∴ when $\dfrac{d^2\theta}{dt^2} = -4$ then $-4 = 10 - 6t$

or $t = 2.33$ seconds

1 If $s = 10 + 50t - 2t^2$, where s metres is the distance travelled in t seconds by a body, what is the velocity of the body after 2 seconds?

2 If $v = 5 + 24t - 3t^2$ where v m/s is the velocity of a body at a time t seconds, what is the acceleration when $t = 3$?

3 A body moves s metres in t seconds where $s = t^3 - 3t^2 - 3t + 8$. Find:

(a) its velocity at the end of 3 seconds
(b) when its velocity is zero
(c) its acceleration at the end of 2 seconds
(d) when its acceleration is zero

4 A body moves s metres in t seconds, where $s = \dfrac{1}{t^2}$. Find the velocity and acceleration after 3 seconds.

5 The distance s metres travelled by a falling body starting from rest after a time t seconds is given by $s = 5t^2$. Find its velocity after 1 second and after 3 seconds.

6 The distance s metres moved by the end of a lever after a time t seconds is given by the formula $s = 6t^2$. Find the velocity of the end of the lever when it has moved a distance $\frac{1}{2}$ metre.

7 The angular displacement θ radians of the spoke of a wheel is given by the expression $\theta = \frac{1}{2}t^4 - t^3$ where t seconds is the time. Find:

(a) the angular velocity after 2 seconds
(b) the angular acceleration after 3 seconds
(c) when the angular acceleration is zero

8 An angular displacement θ radians in time t seconds is given by the equation $\theta = \sin 3t$. Find:

(a) the angular velocity when $t = 1$ second
(b) the smallest positive value of t for which the angular velocity is 2 rad/s
(c) the angular acceleration when $t = 0.5$ seconds
(d) the smallest positive value of t for which the angular acceleration is 9 rad/s^2

9 A mass of 5000 kg moves along a straight line so that the distance s metres travelled in a time t seconds is given by $s = 3t^2 + 2t + 3$. If v m/s is its velocity and m kg is its mass, then its kinetic energy is given by the formula $\frac{1}{2}mv^2$. Find its kinetic energy at a time $t = 0.5$ seconds, remembering that the joule (J) is the unit of energy.

Maximum and Minimum

OBJECTIVES

1 Define the turning point of a graph.

2 Determine the derivative of the function of the graph concerned.

3 Determine the value of x (the independent variable) at the turning points using **1** and **2**.

4 Evaluate y (the dependent variable) corresponding to the values in **3**.

5 Determine the nature of the turning points by consideration of the gradient on either side of the point.

6 Determine and evaluate the second derivative of the function at the turning points.

7 Determine the nature of the turning points by the sign of the second derivative.

8 Solve problems involving maxima and minima relevant to technology.

Turning Points

At the points P and Q (Fig. 12.1) the tangent to the curve is parallel to the x-axis. The points P and Q are called **turning points**. The turning point at P is called a **maximum** turning point and the turning point at Q is called a *minimum* turning point. It will be seen from Fig. 12.1 that the value of y at P is not the greatest value of y nor is the value of y at Q the least. The terms 'maximum' and 'minimum' values apply only in the vicinity of the turning points and not to the values of y in general.

Fig. **12.1**

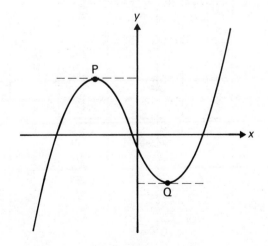

In practical applications, however, we are usually concerned with a specific range of values of x which are dictated by the problem. There is then no difficulty in identifying a particular maximum or minimum within this range of values of x.

Example 12.1

Plot the graph of $y = x^3 - 5x^2 + 2x + 8$ for values of x between -2 and 6. Hence find the maximum and minimum values of y.

To plot the graph we draw up a table in the usual way:

x	-2	-1	0	1	2	3	4	5	6
$y = x^3 - 5x^2 + 2x + 8$	-24	0	8	6	0	-4	0	18	56

The graph is shown in Fig. 12.2. The maximum value occurs at the point P where the tangent to the curve is parallel to the x-axis. The

Fig. **12.2**

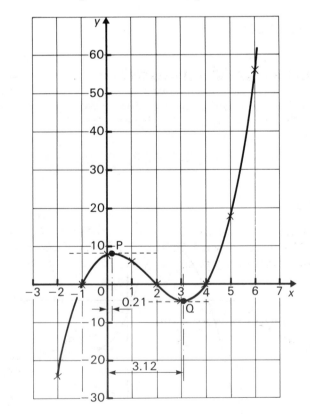

minimum value occurs at the point Q where again the tangent to the curve is parallel to the x-axis. From the graph the maximum value of y is 8.21 and the minimum value of y is −4.06.

Notice that the value of y at P is not the greatest value of y, nor is the value of y at Q the least. However, the values of y at P and Q are called the 'maximum' and 'minimum' values of y respectively.

It is not always convenient to draw the full graph to find the turning points as in the previous example. At a turning point the tangent to the curve is parallel to the x-axis (Fig. 12.3) and hence the gradient of the curve is zero, i.e. $\frac{dy}{dx} = 0$. Using this fact enables us to find the values of x at which the turning points occur.

Fig. **12.3**

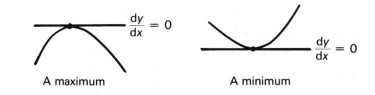

A maximum   A minimum

Maximum or Minimum?

It is often necessary to determine whether each point is a maximum or a minimum.

Two methods of testing are as follows.

Method 1

Consider the gradients of the curve on either side of the turning point. Fig. 12.4 shows how the gradient (or slope) of curve changes in the vicinity of a turning point.

Fig. **12.4**

A maximum   A maximum

Method 2

Find the value of $\dfrac{d^2y}{dx^2}$ at the turning point.

If it is positive then the turning point is a minimum, and if it is negative, then the turning point is a maximum.

If the original expression can be differentiated twice and the expression for $\dfrac{d^2y}{dx^2}$ obtained without too much difficulty, then the second method is generally used.

Example 12.2

Find the maximum and minimum values of y given that

$$y = x^3 + 3x^2 - 9x + 6$$

We have
$$y = x^3 + 3x^2 - 9x + 6.$$

$\therefore$
$$\frac{dy}{dx} = 3x^2 + 6x - 9$$

and
$$\frac{d^2y}{dx^2} = 6x + 6$$

At a turning point
$$\frac{dy}{dx} = 0$$

$\therefore \qquad 3x^2 + 6x - 9 = 0$

$\therefore \qquad x^2 + 2x - 3 = 0 \quad$ by dividing through by 3

$\therefore \qquad (x - 1)(x + 3) = 0$

$\therefore$ either $\qquad x - 1 = 0 \quad$ or $\quad x + 3 = 0$

$\therefore$ either $\qquad x = 1 \quad$ or $\qquad x = -3$

Test for maximum or minimum:

From the above we have

$$\frac{d^2y}{dx^2} = 6x + 6$$

$\therefore$ at the point where $x = 1$, $\dfrac{d^2y}{dx^2} = 6(1) + 6 = +12$

This is positive and hence the turning point at $x = 1$ is a minimum.

The minimum value of y may be found by substituting $x = 1$ into the given equation. Hence

$$y_{min} = (1)^3 + 3(1)^2 - 9(1) + 6 = 1$$

At the point where $x = -3$, $\dfrac{d^2y}{dx^2} = 6(-3) + 6 = -12.$

This is negative and hence at $x = -3$ there is a maximum turning point. The maximum value of y may be found by substituting $x = -3$ into the given equation. Hence

$$y_{max} = (-3)^3 + 3(-3)^2 - 9(-3) + 6 = 3$$

Maximum or minimum using method 1:

At the turning point where $x = 1$, we know that

$$\frac{dy}{dx} = 0$$

i.e. there is zero slope

and using a value of x slightly less than 1, say $x = 0.5$, gives

$$\frac{dy}{dx} = 3(0.5)^2 + 6(0.5) - 9 = -5.25$$

i.e. there is a negative slope

and using a value of x slightly greater than 1, say $x = 1.5$, gives

$$\frac{dy}{dx} = 3(1.5)^2 + 6(1.5) - 9 = +6.75$$

i.e. there is a positive slope

These results are best shown by means of a diagram (Fig. 12.5) which indicates clearly that when $x = 1$ we have a minimum.

Fig. **12.5**

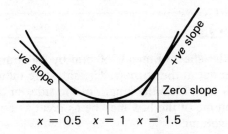

-ve slope
+ve slope
Zero slope
$x = 0.5$ $x = 1$ $x = 1.5$

Now at the turning point where $x = -3$ we know that

$$\frac{dy}{dx} = 0,$$

i.e. there is zero slope

and using a value of x slightly less than -3, say $x = -3.5$, gives

$$\frac{dy}{dx} = 3(-3.5)^2 + 6(-3.5) - 9 = +6.75$$

i.e. there is a positive slope

and using a value of x slightly greater than -3, say $x = -2.5$, gives

$$\frac{dy}{dx} = 3(-2.5)^2 + 6(-2.5) - 9 = -5.25$$

i.e. there is a negative slope

Fig. 12.6 indicates that when $x = -3$ we have a maximum turning point.

Fig. **12.6**

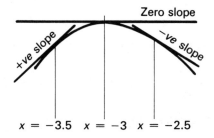

$x = -3.5$ $x = -3$ $x = -2.5$

Applications in Technology

There are many applications in technology which involve the finding of maxima and minima. The first step is to construct an equation connecting the quantity for which a maximum or minimum is required in terms of another variable. A diagram representing the problem may help in the formation of this initial equation.

Example 12.3

A rectangular sheet of metal 360 mm by 240 mm has four equal squares cut out at the corners. The sides are then turned up to form a rectangular box. Find the length of the sides of the squares cut out so that the volume of the box may be as great as possible, and find this maximum volume.

Let the length of the side of each cut-away square be x mm as shown in Fig. 12.7.

Fig. **12.7**

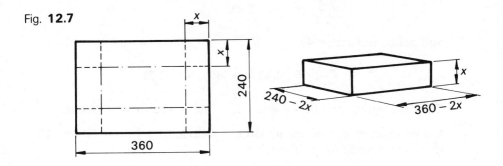

Hence the volume is

$$V = x(240 - 2x)(360 - 2x)$$
$$= 4x^3 - 1200x^2 + 86\,400x$$

$\therefore$ $$\frac{dV}{dx} = 12x^2 - 2400x + 86\,400$$

and $$\frac{d^2V}{dx^2} = 24x - 2400$$

At a turning point $$\frac{dV}{dx} = 0$$

$\therefore$ $$12x^2 - 2400x + 86\,400 = 0$$

or $x^2 - 200x + 7200 = 0$ by dividing through by 12

Now this is a quadratic equation which does not factorise so we shall have to solve it using the formula for the standard quadratic

$$ax^2 + bx + c = 0, \quad \text{which gives} \quad x = \frac{-b \pm \sqrt{b^2 - 4ac}}{2a}.$$

Hence the solution of our equation is

$$x = \frac{-(-200) \pm \sqrt{(-200)^2 - 4 \times 1 \times 7200}}{2 \times 1}$$

$\therefore$ either $x = 152.9$ or $x = 47.1$

However, from the physical sizes of the sheet, it is not possible for x to be 152.9 mm (since one side is only 240 mm long) so we reject this solution. Hence $x = 47.1$ mm.

Test for maximum or minimum:

From the above we have

$$\frac{d^2V}{dx^2} = 24x - 2400$$

and hence when $x = 47.1$

$$\frac{d^2V}{dx^2} = 24(47.1) - 2400 = -1270$$

This is negative, and hence V is a maximum when $x = 47.1$ mm.

It only remains to find the maximum volume by substituting $x = 47.1$ into the equation for V. Therefore

$$V_{max} = 47.1(240 - 2 \times 47.1)(360 - 2 \times 47.1)$$
$$= 1.825 \times 10^6 \text{ mm}^3$$

Example 12.4

A cylinder with an open top has a capacity of 2 m^3 and is made from sheet metal. Neglecting any overlaps at the joints find the dimensions of the cylinder so that the amount of sheet steel used is a minimum.

Let the height of a cylinder be h metres and the radius of the base be r metres as shown in Fig. 12.8.

Fig. **12.8**

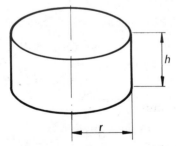

Now   the total area of metal = area of base + area of curved side

$$A = \pi r^2 + 2\pi r h$$

We cannot proceed to differentiate as there are two variables on the right-hand side of the equation. It is possible, however, to find a connection between r and h using the fact that the volume is 2 m^3.

Now $\qquad$ Volume of a cylinder $= \pi r^2 h$

$\therefore \qquad\qquad\qquad\qquad 2 = \pi r^2 h$

from which $\qquad\qquad\qquad\qquad h = \dfrac{2}{\pi r^2}$

We may now substitute for h in the equation for A.

$\therefore \qquad\qquad\qquad A = \pi r^2 + 2\pi r \left(\dfrac{2}{\pi r^2} \right)$

$$= \pi r^2 + \frac{4}{r}$$

$$= \pi r^2 + 4r^{-1}$$

$\therefore \qquad\qquad\qquad \dfrac{\mathrm{d}A}{\mathrm{d}r} = 2\pi r - 4r^{-2}$

and $\qquad\qquad\qquad \dfrac{\mathrm{d}^2 A}{\mathrm{d}r^2} = 2\pi + 8r^{-3}$

Now for a turning point

$$\frac{\mathrm{d}A}{\mathrm{d}r} = 0$$

or $\qquad\qquad\qquad 2\pi r - 4r^{-2} = 0$

$\therefore \qquad\qquad\qquad 2\pi r - \dfrac{4}{r^2} = 0$

$\therefore \qquad\qquad\qquad 2\pi r = \dfrac{4}{r^2}$

$\therefore \qquad\qquad\qquad r^3 = \dfrac{2}{\pi} = 0.637$

$$r = \sqrt[3]{0.637} = 0.860$$

To test for a minimum:

From above we have $\qquad \dfrac{\mathrm{d}^2 A}{\mathrm{d}r^2} = 2\pi + 8r^{-3}$

$$= 2\pi + \frac{8}{r^3}$$

We do not need to do any further calculation here as this expression must be positive for all positive values of r. Hence $r = 0.86$ makes A a minimum.

We may find the corresponding value of h by substituting $r = 0.86$ into the equation found previously for h in terms of r.

$$h = \frac{2}{\pi(0.86)^2} = 0.86$$

Hence for the minimum amount of metal to be used the radius is 0.86 m and the height is 0.86 m.

EXERCISE

12·1

1 Find the maximum and minimum values of:
   (a) $y = 2x^3 - 3x^2 - 12x + 4$
   (b) $y = x^3 - 3x^2 + 4$
   (c) $y = 6x^2 + x^3$

2 Given that $y = 60x + 3x^2 - 4x^3$, calculate:
   (a) the gradient of the tangent to the curve of y at the point where $x = 1$
   (b) the value of x for which y has its maximum value
   (c) the value of x for which y has its minimum value

3 Calculate the coordinates of the points on the curve

   $$y = x^3 - 3x^2 - 9x + 12$$

   at each of which the tangent to the curve is parallel to the x-axis.

4 A curve has the equation $y = 8 + 2x - x^2$. Find:
   (a) the value of x for which the gradient of the curve is 6
   (b) the value of x which gives the maximum value of y
   (c) the maximum value of y

5 The curve $y = 2x^2 + \dfrac{k}{x}$ has a gradient of 5 when $x = 2$.
   Calculate:
   (a) the value of k
   (b) the minimum value of y

6 A rectangular sheet of metal measuring 120 mm by 75 mm has equal squares of side x cut from each of the corners. The remaining flaps are then folded upwards to form an open box. Prove that the volume of the box is given by

   $$V = 9000x - 390x^2 + 4x^3$$

   Find the value of x such that the volume is a maximum.

7 An open rectangular tank of height h metres with a square base of side x metres is to be constructed so that it has a capacity of 500 cubic metres. Show that the inner surface area of the four walls and the base will be $\dfrac{2000}{x} + x^2$ square metres. Find the value of x for this expression to be a minimum.

8 The volume of a cone is given by the formula $V = \frac{1}{3}\pi r^2 h$, where h is the height of the cone and r its radius. If $h = 6 - r$, calculate the value of r for which the volume is a maximum.

9 A box without a lid has a square base of side x mm and rectangular sides of height h mm. It is made from $10\,800$ mm^2 of sheet metal of negligible thickness. Prove that $h = \dfrac{10\,800 - x^2}{4x}$ and that the volume of the box is $(2700x - \frac{1}{4}x^3)$. Hence calculate the maximum volume of the box.

10 A cylindrical tank, with an open top, is to be made to hold 300 cubic metres of liquid. Find the dimensions of the tank so that its surface area shall be a minimum.

11 A cooling tank is to be made with the trapezoidal section shown:

Its cross-sectional area is to be $300\,000$ mm^2. Show that the width of material needed to form, from one sheet, the bottom and folded-up sides is $w = \dfrac{300\,000}{h} + 1.828h$.

Hence find the height h of the tank so that the width of material needed is a minimum.

12 A cylindrical cup is to be drawn from a disc of metal of 50 mm diameter. Assuming that the surface area of the cup is the same as that of the disc find the dimensions of the cup so that its volume is a maximum.

13 A lever weighing 12 N per m run of its length is as shown:

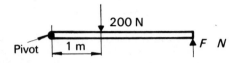

Find the length of the lever so that the force F shall be a minimum.

14 The cost per hour of running a certain machine is

$$C = 1.20 + 0.06\,N^3$$

where N is the number of components produced per hour. Find the most economical value of N if 1000 components are to be produced.

15 A rectangle is inscribed in a circle of 120 mm diameter. Show that the rectangle having the largest area is a square, and find the length of its side.

16 The efficiency of a steam turbine is given by

$$\eta = 4(n\rho \cos \alpha - n^2\rho^2)$$

where n and α are constants. Find the maximum value of η.

Partial Fractions

OBJECTIVES

1 Express compound fractions of the type

$$\frac{f(x)}{(x-a)(x-b)(x-c)}, \quad \frac{f(x)}{(x-a)^2(x-b)} \quad \text{or} \quad \frac{f(x)}{(x^2+a^2)(x-b)}$$

in their appropriate partial fraction forms, providing that f(x) is a polynomial of degree less than three.

2 Obtain partial fractions for a compound fraction whose numerator is of the same degree as the denominator.

Compound Fractions and Partial Fractions

Suppose we wish to express $\dfrac{2}{(x+1)} + \dfrac{3x+6}{(x^2+2)}$ as a single fraction.

The LCM of the denominator is $(x+1)(x^2+2)$ and so

$$\frac{2}{(x+1)} + \frac{3x+6}{(x^2+2)} = \frac{2(x^2+2)+(3x+6)(x+1)}{(x+1)(x^2+2)}$$

$$= \frac{2x^2+4+3x^2+3x+6x+6}{(x+1)(x^2+2)} = \frac{5x^2+9x+10}{(x+1)(x^2+2)}$$

It is often necessary to reverse the above procedure and express

$$\frac{5x^2+9x+10}{(x+1)(x^2+2)} \quad \text{as} \quad \frac{2}{(x+1)} + \frac{3x+6}{(x^2+2)}$$

Now $\dfrac{5x^2+9x+10}{(x+1)(x^2+2)}$ is known as a **compound fraction**

while $\dfrac{2}{(x+1)}$ and $\dfrac{3x+6}{(x^2+2)}$ are known as **partial fractions**.

We see that in each of these fractions the numerator is of a smaller degree than the denominator. Remember that the degree of an expression depends on the highest power of the variable — thus in the second partial fraction the numerator $3x+6$ contains the first power of x and is of the first degree while the denominator x^2+2 contains the second power of x and is of the second degree.

175

How to Find Partial Fractions

Example 13.1

Express $\dfrac{x+2}{(1+x)(2-x)}$ in terms of partial fractions.

The numerator $x+2$ is of the first degree while the denominator $(1+x)(2-x) = 2+x-x^2$ is of the second degree and thus we shall be able to find suitable partial fractions.

Now only experience will enable you to decide the form of the partial fractions, but we must remember that in each case the numerator must be of a smaller degree than the denominator. If A and B are constants (i.e. numbers), then let

$$\frac{x+2}{(1+x)(2-x)} \equiv \frac{A}{(1+x)} + \frac{B}{(2-x)}$$

or

$$\frac{x+2}{(1+x)(2-x)} \equiv \frac{A(2-x)+B(1+x)}{(1+x)(2-x)}$$

You will note that the identity sign is used, which means that the expression is true for any value of the variable x. Comparing both sides of the identity we see that the denominators are the same and so we must arrange for the numerators to be the same also.

Thus $\qquad x+2 = A(2-x)+B(1+x)$

Now by choosing suitable values of x it is possible to find the values of the constants A and B.

When $x = 2$

$$2+2 = A(2-2)+B(1+2)$$

$\therefore \qquad 4 = A\times 0 + B\times 3$

from which $\qquad B = \tfrac{4}{3}$

Also when $x = -1$

$$-1+2 = A[2-(-1)]+B[1+(-1)]$$

$\therefore \qquad 1 = A\times 3 + B\times 0$

from which $\qquad A = \tfrac{1}{3}$

Thus

$$\frac{x+2}{(1+x)(2-x)} \equiv \frac{\tfrac{1}{3}}{(1+x)} + \frac{\tfrac{4}{3}}{(2-x)} \equiv \frac{1}{3(1+x)} + \frac{4}{3(2-x)}$$

Example 13.2

Express $\dfrac{2x^2 + 7x - 17}{(x-1)(x-2)(x+3)}$ in terms of partial fractions.

This is possible since the numerator is of the second degree, which is one less than the degree of the denominator.

Let $\qquad \dfrac{2x^2 + 7x - 17}{(x-1)(x-2)(x+3)} \equiv \dfrac{A}{(x-1)} + \dfrac{B}{(x-2)} + \dfrac{C}{(x+3)}$

or $\qquad \dfrac{2x^2 + 7x - 17}{(x-1)(x-2)(x+3)} \equiv \dfrac{\begin{array}{c}A(x-2)(x+3) + B(x-1)(x+3) \\ + C(x-1)(x-2)\end{array}}{(x-1)(x-2)(x+3)}$

The denominators are the same, and we must arrange that the numerators are identical.

Thus $\quad 2x^2 + 7x - 17 \equiv A(x-2)(x+3) + B(x-1)(x+3) + C(x-1)(x-2)$

When $x = 1$

$\qquad 2(1)^2 + 7(1) - 17 = A(1-2)(1+3) + B(1-1)(1+3) + C(1-1)(1-2)$

$\therefore \qquad 2 + 7 - 17 = A(-1)(4) + B(0)(4) + C(0)(-1)$

from which $\qquad A = 2$

Also when $x = -3$

$2(-3)^2 + 7(-3) - 17 = \quad A(-3-2)(-3+3) + B(-3-1)(-3+3) \\ \qquad\qquad\qquad\qquad\qquad + C(-3-1)(-3-2)$

$\therefore \qquad 18 - 21 - 17 = A(-5)(0) + B(-4)(0) + C(-4)(-5)$

from which $\qquad C = -1$

Also when $x = 2$

$\qquad 2(2)^2 + 7(2) - 17 = A(2-2)(2+3) + B(2-1)(2+3) + C(2-1)(2-2)$

$\therefore \qquad 8 + 14 - 17 = A(0)(5) + B(1)(5) + C(1)(0)$

from which $\qquad B = 1$

Thus

$$\dfrac{2x^2 + 7x - 17}{(x-1)(x-2)(x+3)} \equiv \dfrac{2}{(x-1)} + \dfrac{1}{(x-2)} + \dfrac{1}{(x+3)}$$

Example 13.3

Express $\dfrac{6x^2 + 9x + 1}{(x + 1)^2(x - 1)}$ in terms of partial fractions.

This is possible since the numerator is of the second degree which is smaller than the denominator which is the third degree. Now when the denominator of the given compound fraction contains a factor to a power — here $(x + 1)^2$ — then the partial fractions must be of the form shown.

Let $\dfrac{6x^2 + 9x + 1}{(x + 1)^2(x - 1)} \equiv \dfrac{A}{(x + 1)^2} + \dfrac{B}{(x + 1)} + \dfrac{C}{(x - 1)}$

or $\dfrac{6x^2 + 9x + 1}{(x + 1)^2(x - 1)} \equiv \dfrac{A(x - 1) + B(x + 1)(x - 1) + C(x + 1)^2}{(x + 1)^2(x - 1)}$

The denominators are the same, and we must arrange for the numerators to be identical.

Thus $6x^2 + 9x + 1 \equiv A(x - 1) + B(x + 1)(x - 1) + C(x + 1)^2$

When $x = 1$

$$6(1)^2 + 9(1) + 1 = A(1 - 1) + B(1 + 1)(1 - 1) + C(1 + 1)^2$$

$\therefore$ $6 + 9 + 1 = A(0) + B(2)(0) + C(2)^2$

from which $C = 4$

and when $x = -1$

$$6(-1)^2 + 9(-1) + 1 = A(-1 - 1) + B(-1 + 1)(-1 - 1)$$
$$+ C(-1 + 1)^2$$

$\therefore$ $6 - 9 + 1 = A(-2) + B(0)(-2) + C(0)^2$

from which $A = 1$

We now use the facts that $A = 1$ and $C = 4$ and choose any value for x other than 1 or -1 used previously. For simplicity we use $x = 0$.

Then

$$6(0)^2 + 9(0) + 1 = 1(0 - 1) + B(0 + 1)(0 - 1) + 4(0 + 1)^2$$

from which $B = 2$

Thus

$$\frac{6x^2 + 9x + 1}{(x + 1)^2(x - 1)} \equiv \frac{1}{(x + 1)^2} + \frac{2}{(x + 1)} + \frac{4}{(x - 1)}$$

Example 13.4

Express $\dfrac{3x^2 + 5x + 1}{(x^2 + 4)(x + 3)}$ in terms of partial fractions.

Again the problem is possible since the numerator is of a lesser degree than the denominator. Here again the form of the partial fractions is known by experience and you should remember this.

Let $\qquad \dfrac{3x^2 + 5x + 1}{(x^2 + 4)(x + 3)} = \dfrac{Ax + B}{(x^2 + 4)} + \dfrac{C}{(x + 3)}$

or $\qquad \dfrac{3x^2 + 5x + 1}{(x^2 + 4)(x + 3)} = \dfrac{(Ax + B)(x + 3) + C(x^2 + 4)}{(x^2 + 4)(x + 3)}$

The denominators are the same and so we must equate the numerators.

Thus $\qquad 3x^2 + 5x + 1 \equiv (Ax + B)(x + 3) + C(x^2 + 4)$

When $x = -3$

$$3(-3)^2 + 5(-3) + 1 = [A(-3) + B](-3 + 3) + C[(-3)^2 + 4]$$

$\therefore \qquad 27 - 15 + 1 = (-3A + B)(0) + C(9 + 4)$

from which $\qquad C = 1$

It is not possible to find the values of the remaining constants A and B by substituting specific values of x — try it for yourself. An alternative method is used in which we equate 'like' terms on the left and right-hand sides of the identity. This will necessitate multiplying out the right-hand side.

Hence $\qquad 3x^2 + 5x + 1 \equiv (Ax + B)(x + 3) + C(x^2 + 4)$

will become $\quad 3x^2 + 5x + 1 \equiv Ax^2 + 3Ax + Bx + 3B + Cx^2 + 4C$

and putting $C = 1$ gives

$$3x^2 + 5x + 1 = (A + 1)x^2 + (3A + B)x + (3B + 4)$$

Equating coefficients of x^2 gives

$$3 = A + 1$$

$\therefore \qquad A = 2$

and equating coefficients of x gives

$$5 = (3A + B)$$

But $A = 2$, thus $\qquad 5 = (3 \times 2) + B$

from which $\qquad B = -1$

Thus $\dfrac{3x^2 + 5x + 1}{(x^2 + 4)(x + 3)} \equiv \dfrac{2x - 1}{(x^2 + 4)} + \dfrac{1}{(x + 3)}$

Example 13.5

Express $\dfrac{x^2 + 5x + 5}{(x + 1)(x + 2)}$ in terms of partial fractions.

We cannot proceed here as the degree of the numerator is the same as that of the denominator. We must, therefore, divide the denominator into the numerator. The denominator $(x + 1)(x + 2)$ when multiplied out is $x^2 + 3x + 2$.

Thus

$$x^2 + 3x + 2) \overline{\smash{)}x^2 + 5x + 5} \;(\; 1$$
$$\underline{x^2 + 3x + 2}$$
$$2x + 3$$

Hence the result is 1 together with a remainder of $2x + 3$

or $\dfrac{x^2 + 5x + 5}{(x + 1)(x + 2)} \equiv 1 + \dfrac{2x + 3}{(x + 1)(x + 2)}$

The compound fraction on the right-hand side of the identity has a numerator of lesser degree than the denominator and so may be expressed in terms of partial fractions in a manner similar to that used in Example 13.1. You may care to check this yourself.

Hence $\dfrac{x^2 + 5x + 5}{(x + 1)(x + 2)} \equiv 1 + \dfrac{1}{(x + 1)} + \dfrac{1}{(x + 2)}$

EXERCISE

13·1   Express in terms of partial fractions:

1 $\dfrac{5x - 3}{(x - 3)(x + 3)}$ 2 $\dfrac{x}{(x - 1)(x - 2)}$

3 $\dfrac{5x - 7}{(x + 1)(2x - 1)}$ 4 $\dfrac{x + 7}{x^2 + 3x + 2}$

5 $\dfrac{9x^2 + 34x + 29}{(x + 1)(x + 2)(x + 3)}$ 6 $\dfrac{3x^2 - 7x + 2}{(2x - 1)(x + 1)(x - 1)}$

7 $\dfrac{6 - 15x + 7x^2}{2x(1 - x)(2 - 3x)}$ 8 $\dfrac{2x + 4}{(1 + x)(x - 1)(2x + 1)}$

9 $\dfrac{7x - 2x^2}{(2-x)^2(1+x)}$

10 $\dfrac{10 - 2x - 3x^2}{(x-1)^2(2x+3)}$

11 $\dfrac{x^2 + 3}{(x+1)(x-1)^2}$

12 $\dfrac{-(2+3x)}{x(1+x)^2}$

13 $\dfrac{3x^2 + 3x + 2}{(x^2+1)(x+1)}$

14 $\dfrac{2x^2 + 2x - 7}{(x-1)(2+x^2)}$

15 $\dfrac{10 + 5x - 8x^2}{(2-3x^2)(4+x)}$

16 $\dfrac{6x^2 + 8x + 16}{(x+2)(x^2+4)}$

17 $\dfrac{x^2 + 4x - 2}{(x-1)(x+2)}$

18 $\dfrac{x(2x-3)}{(2x-1)(x-1)}$

19 $\dfrac{x^2 - 3x - 2}{(x-1)(x+1)}$

20 $\dfrac{3(x^2 - x - 1)}{(x+1)(x-2)}$

Integration

OBJECTIVES

1 Determine indefinite integrals of functions involving sin ax, cos ax and e^{ax}.
2 Evaluate the definite integrals involving sin ax, cos ax and e^{ax}.
3 Define the mean and root mean square values of functions over a given range.
4 Evaluate the mean and root mean square values of simple periodic functions.

Introduction

The table of differential coefficients on p. 144 shows that by differentiating sin ax with respect to x we obtain $a \cos ax$. Hence by differentiating $\dfrac{1}{a} \sin ax$ with respect to x we obtain $\cos ax$.

Now since integration is the reverse of differentiation, it follows that by integrating cos ax with respect to x we obtain $\dfrac{1}{a} \sin ax$.

Now by making small modifications similar to the one just described, we may rewrite the table, showing integrals of the more common functions:

y	$\int y \, dx$
ax^n	$\dfrac{a}{n+1} x^{n+1}$
$\sin ax$	$-\dfrac{1}{a} \cos ax$
$\cos ax$	$\dfrac{1}{a} \sin ax$
$\sec^2 x$	$\tan x$
$\dfrac{1}{x}$	$\log_e x$
e^{ax}	$\dfrac{1}{a} e^{ax}$

Indefinite Integrals

An **indefinite** integral is an integral *without limits* and the solution must therefore contain a constant of integration.

Example 14.1

Find $\int (x^2 + 2x - 3)\, dx$.

$$\int (x^2 + 2x - 3)\, dx = \frac{x^3}{3} + 2\frac{x^2}{2} - 3x + k$$

$$= \frac{x^3}{3} + x^2 - 3x + k$$

Example 14.2

Find $\int (\sin 7\theta + 2 \cos 5\theta)\, d\theta$.

$$\int (\sin 7\theta + 2 \cos 5\theta)\, d\theta = -\tfrac{1}{7} \cos 7\theta + \tfrac{2}{5} \sin 5\theta + k$$

Example 14.3

Find $\int \left(e^{6t} - \dfrac{1}{e^{3t}} \right) dt$.

$$\int \left(e^{6t} - \frac{1}{e^{3t}} \right) dt = \int (e^{6t} - e^{-3t})\, dt$$

$$= \tfrac{1}{6} e^{6t} - \frac{1}{(-3)} e^{-3t} + k$$

$$= \tfrac{1}{6} e^{6t} + \tfrac{1}{3} e^{-3t} + k$$

Definite Integrals

A **definite** integral *has limits*. Square brackets are used to indicate that we have completed the actual integration and will next be substituting the values of the limits.

Example 14.4

Evaluate $\int_2^3 (1 + \cos 2\phi)\, d\phi$.

We should remember that when limits are substituted into
trigonometrical functions they represent radian values (not degrees).

$$\int_2^3 (1 + \cos 2\phi)\, d\phi = \left[\phi + \tfrac{1}{2} \sin 2\phi\right]_2^3$$

$$= \{3 + \tfrac{1}{2} \sin (2 \times 3)\} - \{2 + \tfrac{1}{2} \sin (2 \times 2)\}$$

$$= 3 + \tfrac{1}{2} \sin 6 - 2 - \tfrac{1}{2} \sin 4$$

$$= 1.24$$

Example 14.5

Evaluate $\int_0^1 5(e^{2t} - e^{-2t})\, dt$.

$$\int_0^1 5(e^{2t} - e^{-2t})\, dt = 5 \int_0^1 (e^{2t} - e^{-2t})\, dt$$

$$= 5 \left[\tfrac{1}{2} e^{2t} - \frac{1}{(-2)} e^{-2t}\right]_0^1$$

$$= \tfrac{5}{2}\{(e^{2 \times 1} + e^{-2 \times 1}) - (e^{2 \times 0} + e^{-2 \times 0})\}$$

$$= \tfrac{5}{2}(e^2 + e^{-2} - e^0 - e^0)$$

$$= \tfrac{5}{2}(7.39 + 0.14 - 1 - 1) = 13.8$$

Example 14.6

Evaluate $\int_1^2 \left(\dfrac{1}{x} + \sec^2 x\right) dx$.

$$\int_1^2 \left(\frac{1}{x} + \sec^2 x\right) dx = \left[\log_e x + \tan x\right]_1^2$$

$$= (\log_e 2 + \tan 2) - (\log_e 1 + \tan 1)$$

$$= 0.693 + (-2.185) - 0 - 1.557$$

$$= -3.05$$

EXERCISE

Find:

1 $\int \left(5x^2 + 2x - \dfrac{4}{x^2}\right) dx$

2 $\int \left(\sqrt{x} + \dfrac{1}{\sqrt{x}}\right) dx$

3 $\int \sin \dfrac{x}{3} \, dx$

4 $\int 5 \cos 3\theta \, d\theta$

5 $\int (1 + \sin \tfrac{2}{3}\phi) \, d\phi$

6 $\int \left(\cos \dfrac{\theta}{2} - \sin \dfrac{3\theta}{2}\right) d\theta$

7 $\int (2t + \sin 2t) \, dt$

8 $\int e^{3x} \, dx$

9 $\int e^{-0.5u} \, du$

10 $\int (3e^{2t} - 2e^t) \, dt$

11 $\int (e^{-x/2} + e^{3x/2}) \, dx$

12 $\int \left(\sec^2 x + \dfrac{1}{x}\right) dx$

Evaluate:

13 $\displaystyle\int_1^2 (x^3 + 4) \, dx$

14 $\displaystyle\int_0^2 \left(\dfrac{x^2 + x^3}{x}\right) dx$

15 $\displaystyle\int_0^1 \dfrac{2 \cos x}{3} \, dx$

16 $\displaystyle\int_0^{\pi/2} 3 \sin 4\phi \, d\phi$

17 $\displaystyle\int_{\pi/6}^{\pi/3} 2 \sin \dfrac{2t}{3} \, dt$

18 $\displaystyle\int_{-\pi/2}^{\pi/2} \sin 2\theta \, d\theta$

19 $\displaystyle\int_{\pi/2}^{\pi} \cos \dfrac{\phi}{2} \, d\phi$

20 $\displaystyle\int_{-0.2}^{0.5} (1 + 0.6 \cos 0.2\theta) \, d\theta$

21 $\displaystyle\int_0^{\pi} (\sin x - \sin 3x) \, dx$

22 $\displaystyle\int_0^1 e^x \, dx$

23 $\displaystyle\int_1^2 e^{-2t} \, dt$

24 $\displaystyle\int_{0.5}^1 \left(e^{\theta/3} - \dfrac{1}{e^{\theta/3}}\right) d\theta$

25 $\displaystyle\int_1^2 (1 + 2e^{0.3v}) \, dv$

26 $\displaystyle\int_2^3 4(e^u + e^{-u}) \, du$

27 $\displaystyle\int_0^3 2e^{-0.4t}\,dt$

28 $\displaystyle\int_{-1}^1 \frac{4}{5e^{1.4x}}\,dx$

29 $\displaystyle\int_2^3 \frac{2}{x}\,dx$

30 $\displaystyle\int_{\pi/4}^{\pi/3} \sec^2\theta\,d\theta$

Area Under a Curve

Suppose that we wish to find the shaded area shown in Fig. 14.1.
P, whose coordinates are (x, y), is a point on the curve.

Fig. **14.1**

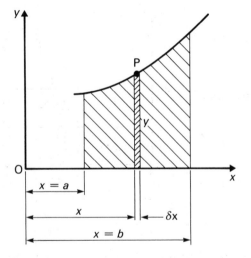

Let us now draw, below P, a vertical strip whose width δx is very small.
Since the width of the strip is very small we may consider the strip to
be a rectangle with height y. Hence the area of the strip is
approximately $y \times \delta x$. Such a strip is called an elementary strip and we
will consider that the shaded area is made up from many elementary
strips. Hence the required area is the sum of all the elementary strip
areas between the values $x = a$ and $x = b$. In mathematical notation
this may be stated as:

$$\text{Area} = \sum_{x=a}^{x=b} y \times \delta x \qquad \text{approximately}$$

The process of integration may be considered to sum up an infinite number of elementary strips and hence gives an exact result.

$\therefore$ $$\text{Area} = \int_a^b y \, dx \quad \text{exactly}$$

Example 14.7

Find the area bounded by the curve $y = x^3 + 3$, the x-axis and the lines $x = 1$ and $x = 3$

It is always wise to sketch the graph of the given curve and show the area required together with an elementary strip, as shown in Fig. 14.2.

Fig. **14.2**

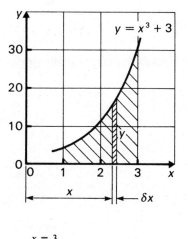

The required area $= \displaystyle\sum_{x=1}^{x=3} y \times \delta x \quad$ approximately

$$= \int_1^3 y \, dx \quad \text{exactly}$$

$$= \int_1^3 (x^3 + 3) \, dx$$

$$= \left[\frac{x^4}{4} + 3x \right]_1^3 = \left(\frac{3^4}{4} + 3 \times 3 \right) - \left(\frac{1^4}{4} + 3 \times 1 \right)$$

$$= 26 \text{ square units}$$

Example 14.8

Find the area under the curve of 2 cos θ between θ = 20° and
θ = 60°.

The curve of 2 cos θ is shown in **Fig. 14.3** from 0° to 90° and the
required area together with an elementary strip.

Fig. **14.3**

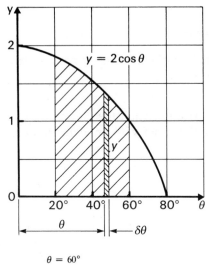

The required area $= \displaystyle\sum_{\theta = 20°}^{\theta = 60°} y \times \delta\theta$ approximately

$$= 2 \int_{20°}^{60°} \cos\theta \, d\theta \quad \text{exactly}$$

$$= 2\left[\sin\theta\right]_{20°}^{60°} = 2\,(\sin 60° - \sin 20°)$$

$$= 1.05 \text{ square units}$$

EXERCISE

1 Find the area between the curve $y = x^3$, the x-axis and the
   lines $x = 5$ and $x = 3$.

2 Find the area between the curve $y = 3 + 2x + 3x^2$, the
   x-axis and the lines $x = 1$ and $x = 4$.

3 Find the area between the curve $y = x^2(2x - 1)$, the x-axis
   and the lines $x = 1$ and $x = 2$.

4 Find the area between the curve $y = \dfrac{1}{x^2}$, the x-axis and
   the lines $x = 1$ and $x = 3$.

5 Find the area between the curve $y = 5x - x^3$, the x-axis and the lines $x = 1$ and $x = 2$.

6 Evaluate the integral $\int_0^{2\pi} \sin \theta \, d\theta$ and explain the result with reference to a sketched graph.

7 Find the area under the curve $2 \sin \theta + 3 \cos \theta$ between $\theta = 0$ and $\theta = \pi$ radians.

8 Find the area under the curve of $y = \sin \phi$ between $\phi = 0$ and $\phi = \pi$ radians.

9 Find the area under the curve $y = e^x$ between the coordinates $x = 0$ and $x = 2$.

10 Find the area under the curve $y = 5e^x$ from $x = -0.5$ to $x = +0.5$.

Mean Value

The mean (or average) value or height of a curve is often of importance.

$$\text{The mean value} = \frac{\text{Area under the curve}}{\text{Length of the base}}$$

A type of graph which is met frequently in technology is a waveform.

A waveform is a graph which repeats indefinitely. A sine curve (Fig. 14.4) is an example of a waveform.

Fig. **14.4**

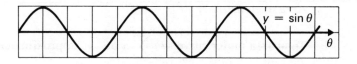

A portion of the graph which shows the complete shape of the waveform without any repetition is called a **cycle**.

In the case of the curve of sin θ the portion of the graph over one cycle is said to be a **full wave** (Fig. 14.5), and over half of one cycle is said to be a **half wave** (Fig. 14.6).

Fig. **14.5**

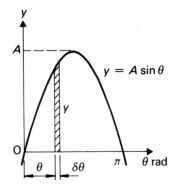

1 cycle

Fig. **14.6**

$\frac{1}{2}$ cycle

Example 14.9

Find the mean value of $A \sin \theta$ for:

(a) a half wave   (b) a full wave

(a) A half wave of $A \sin \theta$ occurs over a range from $\theta = 0$ rad to $\theta = \pi$ rad as shown in Fig. 14.7.

Fig. **14.7**

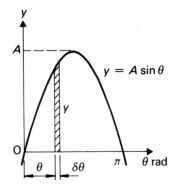

$y = A \sin \theta$

Area of elementary strip $= y \, \delta\theta$

∴   Total area under curve $= \displaystyle\sum_{\theta = 0}^{\theta = \pi} y \, \delta\theta$ approximately

$= \displaystyle\int_{0}^{\pi} y \, d\theta$ exactly

$$= \int_0^\pi A \sin \theta \, d\theta$$

$$= A \int_0^\pi \sin \theta \, d\theta$$

$$= A \left[-\cos \theta \right]_0^\pi$$

$$= A \left\{ (-\cos \pi) - (-\cos 0) \right\}$$

$$= A \left\{ -(-1) - (-1) \right\}$$

$$= 2A \text{ square units}$$

Now   Mean value $= \dfrac{\text{Area under curve}}{\text{Length of base}}$

$$= \frac{2A}{\pi} = 0.637A$$

(b) A full wave of $A \sin \theta$ occurs over a range from $\theta = 0$ rad to $\theta = 2\pi$ rad as shown in Fig. 14.8.

Fig. **14.8**

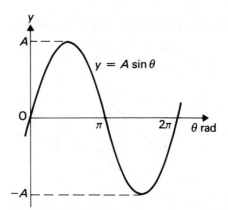

The method used is similar to that in part (a) except that the limits of the integral are now 0 and 2π.

$\therefore$ Total area under curve $= \displaystyle\int_0^{2\pi} A \sin \theta \, d\theta$

$$= A \left[-\cos \theta \right]_0^{2\pi}$$

$$= 0$$

This answer is zero because the area under the second half of the wave is calculated as negative and added to an identical positive area under the first half wave. Thus the mean value is zero.

Example 14.10

A ramp waveform consists of a series of right-angled triangles as shown in Fig. 14.9. Find the mean height of the waveform.

Fig. **14.9**

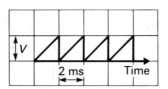

The triangular area under the graph is $\frac{1}{2} \times 2 \times V = V$. However, it is instructive to find the area by integration, as a similar method is used later for finding root mean squares.

We must first find an equation for the sloping side of the triangle, which is shown set up on v–t axes as shown in Fig. 14.10.

Fig. **14.10**

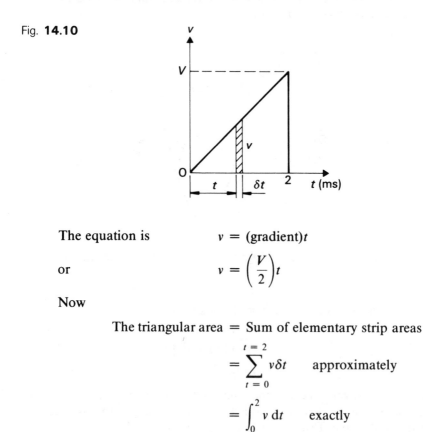

The equation is $v = (\text{gradient})t$

or $v = \left(\dfrac{V}{2}\right)t$

Now

The triangular area = Sum of elementary strip areas

$$= \sum_{t=0}^{t=2} v\delta t \qquad \text{approximately}$$

$$= \int_{0}^{2} v \, dt \qquad \text{exactly}$$

$$= \int_0^2 \left(\frac{V}{2} \right) t \, dt$$

$$= \frac{V}{2} \int_0^2 t \, dt = \frac{V}{2} \left[\frac{t^2}{2} \right]_0^2 = V$$

Now   Mean value $= \dfrac{\text{Area under graph}}{\text{Length of base}}$

$$= \frac{V}{2}$$

Root Mean Square (r.m.s.) Value

In alternating current work the mean value is not of great importance, because we are usually interested in the power produced and this depends on the square of the current or voltage values. In these cases we use the **root mean square** (r.m.s.) value.

The root mean square value $= \sqrt{\text{Average height of the } y^2 \text{ curve}}$

i.e.   r.m.s. $= \sqrt{\dfrac{\text{Area under the } y^2 \text{ curve}}{\text{Length of the base}}}$

Example 14.11

Find the r.m.s. value of $A \sin \theta$ for:

(a) a half wave   (b) a full wave

(a) The method is similar to that used in Example 14.9 except that we use y^2 instead of y in the integral.

Thus

$$\text{Total area under } y^2 \text{ curve} = \int_0^\pi y^2 \, d\theta$$

$$= \int_0^\pi (A \sin \theta)^2 \, d\theta$$

$$= A^2 \int_0^\pi \sin^2\theta \, d\theta$$

However, $\sin^2\theta = \frac{1}{2}(1 - \cos 2\theta)$ as on p. 100.

Therefore

$$\text{Total area under } y^2 \text{ curve} = \frac{A^2}{2} \int_0^\pi (1 - \cos 2\theta) \, d\theta$$

$$= \frac{A^2}{2} \left[\theta - \tfrac{1}{2} \sin 2\theta \right]_0^\pi$$

$$= \frac{A^2}{2} \{ (\pi - \tfrac{1}{2} \sin 2 \times \pi) - (0 - \tfrac{1}{2} \sin 2 \times 0) \}$$

$$= \frac{A^2}{2} \{ \pi - \tfrac{1}{2} \times 0 - 0 + \tfrac{1}{2} \times 0 \}$$

$$= \frac{\pi A^2}{2}$$

Now $$\text{r.m.s. value} = \sqrt{\frac{\text{Area under } y^2 \text{ curve}}{\text{Length of base}}}$$

$$= \sqrt{\frac{\pi A^2 / 2}{\pi}}$$

$$= \frac{A}{\sqrt{2}} = 0.707A$$

(b) For a full wave the working is similar to that in part (a) except that the limits are 0 to 2π rad. The reader may find it useful to verify that the same result of $0.707A$ is obtained.

Similar results are also obtained for cosine waveforms. Thus:

> For sinusoidal waveforms the r.m.s. value is 0.707 of the amplitude or peak value.

Example 14.12

Find the r.m.s. value of the ramp waveform shown in Fig. 14.10.

The method is similar to that used in Example 14.10 except that we use v^2 instead of v in the integral.

Thus

$$\text{Total area under the } v^2 \text{ graph} = \int_0^2 v^2 \, dt$$

$$= \int_0^2 \left\{ \left(\frac{V}{2} \right) t \right\}^2 dt$$

$$= \frac{V^2}{4} \int_0^2 t^2 \, dt$$

$$= \frac{V^2}{4} \left[\frac{t^3}{3} \right]_0^2$$

$$= \frac{V^2}{4} \left(\frac{2^3}{3} - \frac{0^3}{3} \right)$$

$$= \frac{2V^2}{3}$$

Now   r.m.s. value $= \sqrt{\dfrac{\text{Area under } v^2 \text{ graph}}{\text{Length of base}}}$

$$= \sqrt{\frac{2V^2/3}{2}}$$

$$= \frac{V}{\sqrt{3}} = 0.577 V$$

EXERCISE

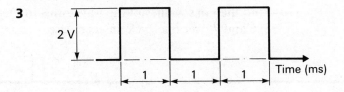
14·3

1 Find the mean value of $V \cos \theta$ for:
   (a) a half cycle
   (b) a full cycle

2 Find the mean value of $6 \sin 2\theta$ for:
   (a) a half cycle
   (b) a full cycle

Find the mean values of the waveforms shown in Questions 3, 4, 5 and 6 over one cycle in each case.

3

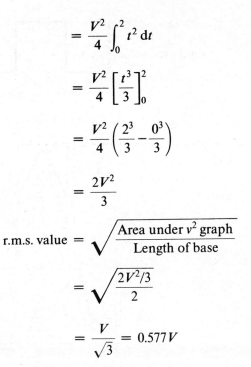

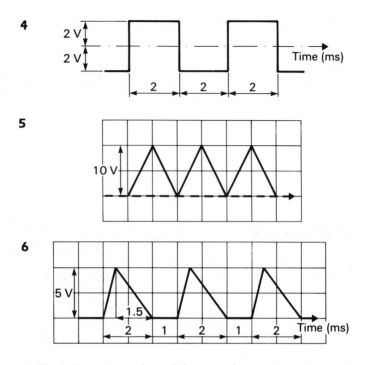

7 Find the mean value of the waveform whose shape for one cycle is as shown.

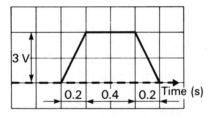

8 Find the r.m.s. value of $V \cos \theta$ over:

(a) a half wave

(b) a full wave

9 Find the r.m.s. value of $6 \sin 2\theta$ over:

(a) a half cycle

(b) a full cycle

10 Find the r.m.s. values of the waveforms in Questions 3, 4, 5, 6 and 7 over one cycle in each case.

Integration by Substitution

Previously we learnt how to differentiate by using a substitution and thus reduce a complicated looking expression to one which was relatively simple. Integration may also be simplified by using a suitable substitution, together with use of the formula

$$\frac{\mathrm{d}y}{\mathrm{d}x} = \frac{\mathrm{d}y}{\mathrm{d}u} \times \frac{\mathrm{d}u}{\mathrm{d}x}$$

Example 14.13

Find $\int \left(\frac{x}{1+x^2}\right) \mathrm{d}x$.

Choosing a suitable substitution is a question of experience, but after working through a number of problems you will find it is not as difficult as it may first appear.

In this instance we will try to simplify the bottom line by putting $u = 1 + x^2$

Hence $\int \left(\frac{x}{1+x^2}\right) \mathrm{d}x = \int \left(\frac{x}{u}\right) \mathrm{d}x$ where $u = 1 + x^2$

$$\therefore \qquad \frac{\mathrm{d}u}{\mathrm{d}x} = 2x$$

$$\therefore \qquad \mathrm{d}x = \frac{\mathrm{d}u}{2x}$$

At this stage we have two variables, x and u, and we shall try to eliminate all the x terms, starting with the $\mathrm{d}x$ term.

Thus $\int \left(\frac{x}{1+x^2}\right) \mathrm{d}x = \int \left(\frac{x}{u}\right)\left(\frac{\mathrm{d}u}{2x}\right)$

$\qquad\qquad\qquad = \frac{1}{2}\int \frac{1}{u}\, \mathrm{d}u$ which is a standard integral.

$\qquad\qquad\qquad = \frac{1}{2}(\log_e u) + c$ note the introduction of the constant c, as this is an indefinite integral.

At this stage all that remains is to express the variable u in terms of x.

Thus $\int \left(\dfrac{x}{1+x^2} \right) \mathrm{d}x = \frac{1}{2} \log_e (1+x^2) + c$

Example 14.14

Find $\int \left(\dfrac{x}{\sqrt{a^2+x^2}} \right) \mathrm{d}x$.

Again we shall try a substitution to simplify the bottom line, putting $u = a^2 + x^2$.

$$\int \left(\frac{x}{\sqrt{a^2+x^2}} \right) \mathrm{d}x = \int \frac{x}{\sqrt{u}}\, \mathrm{d}x \qquad \text{where} \quad u = a^2 + x^2$$

$$= \int \left(\frac{x}{u^{1/2}} \right)\left(\frac{\mathrm{d}u}{2x} \right) \qquad \therefore \qquad \frac{\mathrm{d}u}{\mathrm{d}x} = 2x$$

$$= \frac{1}{2} \int u^{-1/2}\, \mathrm{d}u \qquad \therefore \qquad \mathrm{d}x = \frac{\mathrm{d}u}{2x}$$

$$= \frac{1}{2} \left(\frac{u^{1/2}}{\frac{1}{2}} \right) + c$$

$$= (a^2 + x^2)^{1/2} + c$$

Change of Limits

Definite integrals may also be evaluated using a substitution and changing the limits to values of the new variable. This avoids the necessity of reintroducing the original variable.

Example 14.15

Evaluate $\displaystyle\int_0^1 \left(\frac{1}{(2x+3)^3} \right) \mathrm{d}x$.

This is an integral with respect to variable x, which means that the limits are also values of x. We shall try the substitution $u = 2x + 3$ and this relationship may be used to find values of u which correspond to the limit values of $x = 1$ and $x = 0$.

$$\int_0^1 \left(\frac{1}{(2x+3)^3}\right) dx = \int_3^5 \left(\frac{1}{u^3}\right)\left(\frac{du}{2}\right)$$

where $\quad u = 2x + 3$

$$= \tfrac{1}{2}\int_3^5 (u^{-3})\, du$$

$\therefore \quad \dfrac{du}{dx} = 2$

$$= \tfrac{1}{2}\left[\frac{u^{-2}}{-2}\right]_3^5$$

$\therefore \quad dx = \dfrac{du}{2}$

Change of limits:

$$= (\tfrac{1}{2})(\tfrac{1}{-2})\left[\frac{1}{u^2}\right]_3^5$$

When $x = 1\ \ u = 2(1) + 3 = 5$

$$= -\frac{1}{4}\left(\frac{1}{5^2} - \frac{1}{3^2}\right)$$

When $x = 0\ \ u = 2(0) + 3 = 3$

$$= 0.0178$$

Example 14.16

Evaluate $\displaystyle\int_0^{\pi/2} \sin^3 \theta \cos \theta\, d\theta$.

Here the most complicated expression is $\sin^3 \theta$ or $(\sin \theta)^3$ and we will try putting $u = \sin \theta$, and also change the limits.

$$\int_0^{\pi/2} (\sin \theta)^3 (\cos \theta)\, d\theta = \int_0^1 u^3 (\cos \theta)\left(\frac{du}{\cos \theta}\right)$$

where $\quad u = \sin \theta$

$$= \int_0^1 u^3\, du$$

$\therefore \quad \dfrac{du}{d\theta} = \cos \theta$

$$= \left[\frac{u^4}{4}\right]_0^1$$

$\therefore \quad d\theta = \dfrac{du}{\cos \theta}$

$$= \tfrac{1}{4}(1^4 - 0^4)$$

Limits change:

$$= 0.25$$

when $\quad \theta = \dfrac{\pi}{2}$

$$u = \sin \frac{\pi}{2} = 1$$

and $\quad \theta = 0$

$$u = \sin 0 = 0$$

EXERCISE

14·4

By using suitable substitutions find the following indefinite integrals and evaluate the definite integrals:

1 $\int (2x + 1)^3 \, dx$

2 $\int \sqrt{(x + 3)} \, dx$

3 $\int \frac{1}{(1 - 2x)} \, dx$

4 $\int x(3 - x^2)^3 \, dx$

5 $\int \sin(2\theta - 1) \, d\theta$

6 $\int \frac{dx}{e^{(2x - 1)}}$

7 $\int_4^5 \frac{dx}{(x - 3)}$

8 $\int_2^3 x^2(\sqrt{x^3 + 2}) \, dx$

9 $\int_1^2 e^{(t - 4)} \, dt$

10 $\int_0^{\pi/2} \cos\left(3\theta - \frac{\pi}{2}\right) d\theta$

11 $\int_2^3 x e^{x^2} \, dx$

12 $\int_3^4 \frac{1}{x} \log_e x \, dx$

13 $\int_0^1 \frac{dx}{(3x + 1)^2}$

14 $\int_0^1 x(\sqrt{1 - x^2}) \, dx$

15 $\int_3^4 \frac{x \, dx}{\sqrt[3]{x^2 - 7}}$

16 $\int_0^{\pi} \sin\theta \cos^2\theta \, d\theta$

17 $\int_{-2}^{-1} \frac{e^t}{1 - e^t} \, dt$

18 $\int_0^{\pi} x \cos(x^2) \, dx$

19 $\int_0^{\pi/4} \frac{\cos\theta}{1 - \sin\theta} \, d\theta$

20 $\int_0^2 \frac{x^2 \, dx}{\sqrt{x^3 + 1}}$

21 $\int_1^2 \frac{(x + 1)}{x^2 + 2x + 2} \, dx$

Integration by Trigonometrical Substitution

Some expressions cannot be integrated by direct substitution.

For example consider $\int \frac{dx}{1 + x^2}$.

Suppose that, in order to simplify the bottom line, we try putting

$$u = 1 + x^2$$

from which $\dfrac{du}{dx} = 2x$

or $dx = \dfrac{du}{2x}$

Then $\displaystyle\int \dfrac{dx}{1+x^2}$ becomes $\displaystyle\int \dfrac{1}{u} \times \dfrac{du}{2x}$.

This expression is not suitable for integration as we have two variables, u and x, remaining. If we now substitute for x in terms of u (from $x = \sqrt{1-u}$ then the expression becomes more complicated than the original. Hence we must look for another method.

We may use the trigonometrical identities derived from $\sin^2 \theta + \cos^2 \theta = 1$, which are:

$$\sin^2 \theta = 1 - \cos^2 \theta$$

$$\cos^2 \theta = 1 - \sin^2 \theta$$

$$\sec^2 \theta = 1 + \tan^2 \theta$$

Let us now try the substitution $x = \tan \theta$.

$$\begin{aligned}
\text{Hence} \int \frac{dx}{1+x^2} &= \int \frac{dx}{1+(\tan \theta)^2} \\
&= \int \frac{\sec^2 \theta \, d\theta}{1 + \tan^2 \theta} \\
&= \int \frac{\sec^2 \theta \, d\theta}{\sec^2 \theta} \\
&= \int 1 \, d\theta \\
&= \theta + c
\end{aligned}$$

where $x = \tan \theta$

$\therefore \quad \dfrac{dx}{d\theta} = \sec^2 \theta$

or $dx = \sec^2 \theta \, d\theta$

since $\sec^2 \theta = 1 + \tan^2 \theta$

where c is the constant of integration.

All that remains now is to express θ in terms of x.
Since $x = \tan \theta$,

then $\boxed{\theta = \tan^{-1} x}$

or $\boxed{\theta = \arctan x}$

or $\boxed{\theta = \text{inv} \tan x}$

These are alternative ways of stating, in equation form, that θ is the angle whose tangent is numerically equal to x. Your choice may well depend on the notation used on your scientific calculator but all are equally correct.

Thus $\displaystyle\int \frac{dx}{1+x^2} = \arctan x + c$

Example 14.17

Evaluate $\displaystyle\int_0^3 \frac{dx}{\sqrt{9-x^2}}$.

The denominator may be expressed as $\sqrt{3^2 - x^2}$ and this gives a clue as to how to simplify this expression, since from the trigonometrical identity

$$\cos^2 \theta = 1 - \sin^2 \theta \quad \text{we have} \quad \cos \theta = \sqrt{1 - \sin^2 \theta}$$

which is similar in 'shape'. It is thus worth a try, though success cannot be guaranteed — if it fails we must try something else!

Hence

$$\int_0^3 \frac{dx}{\sqrt{9-x^2}} = \int_0^{\pi/2} \frac{3\cos\theta\, d\theta}{\sqrt{3^2 - (3\sin\theta)^2}}$$

where $x = 3\sin\theta$

$$= \int_0^{\pi/2} \frac{3\cos\theta\, d\theta}{\sqrt{3^2 - 3^2\sin^2\theta}}$$

$\therefore$ $\dfrac{dx}{d\theta} = 3\cos\theta$

or $dx = 3\cos\theta\, d\theta$

$$= \int_0^{\pi/2} \frac{3\cos\theta\, d\theta}{\sqrt{3^2(1 - \sin^2\theta)}}$$

Change of limits:

when $x = 3$, $3 = 3\sin\theta$

$$= \int_0^{\pi/2} \frac{3\cos\theta\, d\theta}{\sqrt{3^2}\sqrt{1 - \sin^2\theta}}$$

or $\sin\theta = 1$

$$= \int_0^{\pi/2} \frac{3\cos\theta\, d\theta}{3\cos\theta}$$

$\therefore$ $\theta = \dfrac{\pi}{2}$

when $x = 0$, $0 = 3\sin\theta$

$$= \int_0^{\pi/2} 1\, d\theta$$

or $\sin\theta = 0$

$$= \left[\theta\right]_0^{\pi/2}$$

$\therefore$ $\theta = 0$

$$= \frac{\pi}{2}$$

Note that angles are usually given in radians, not degrees.

$$= 1.571$$

Example 14.18

Find $\displaystyle\int \frac{dx}{x^2\sqrt{1+x^2}}$.

This does not look very promising, as the only identity resembling the shape of one of the bottom line terms is $\sec^2\theta = 1 + \tan^2\theta$ and there is also the x^2 term. However, we shall try the substitution $x = \tan\theta$ and hope for success.

Hence

$$\int \frac{dx}{x^2\sqrt{1+x^2}} = \int \frac{\sec^2\theta\,d\theta}{\tan^2\theta\sqrt{1+\tan^2\theta}} \qquad \text{where} \qquad x = \tan\theta$$

$$= \int \frac{\sec^2\theta\,d\theta}{\tan^2\theta\sqrt{\sec^2\theta}} \qquad \therefore \qquad \frac{dx}{d\theta} = \sec^2\theta$$

$$= \int \frac{\sec^2\theta\,d\theta}{\tan^2\theta\sec\theta} \qquad \text{or} \qquad dx = \sec^2\theta\,d\theta$$

$$= \int \frac{\sec\theta\,d\theta}{\tan^2\theta}$$

$$= \int \frac{\cos^2\theta\,d\theta}{\cos\theta\sin^2\theta}$$

$$= \int \frac{\cos\theta\,d\theta}{\sin^2\theta}$$

At this stage we shall have to simplify the denominator by making a simple substitution $u = \sin\theta$.

$$\text{Thus} \int \frac{dx}{x^2\sqrt{1+x^2}} = \int \frac{du}{u^2} \qquad \text{where} \qquad u = \sin\theta$$

$$= \int u^{-2}\,du \qquad \therefore \qquad \frac{du}{d\theta} = \cos\theta$$

$$= \frac{u^{-1}}{-1} + c \qquad \text{or} \qquad du = \cos\theta\,d\theta$$

$$= -\frac{1}{u} + c$$

$$= -\frac{1}{\sin\theta} + c$$

Now since $x = \tan\theta$ or $\tan\theta = \dfrac{x}{1}$ we may draw a right-angled triangle and label one angle θ with the side opposite equal to x and the one adjacent equal to 1. Using Pythagoras' theorem the hypotenuse will be $\sqrt{1+x^2}$ as shown in Fig. 14.11.

Fig. **14.11**

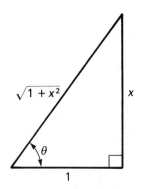

Thus $\sin\theta = \dfrac{x}{\sqrt{1+x^2}}$ and substituting we have:

$$\int \frac{\mathrm{d}x}{x^2\sqrt{1+x^2}} = -\frac{\sqrt{1+x^2}}{x} + c$$

Example 14.19

Evaluate $\displaystyle\int_0^1 \frac{\mathrm{d}\theta}{3\sin\theta + 4\cos\theta}$.

This solution illustrates the use of two more trigonometrical relationships (see Chapter 8):

(1) $a\sin\theta + b\cos\theta = R\sin(\theta + \alpha)$ where $R = \sqrt{a^2 + b^2}$ and $\tan\alpha = \dfrac{b}{a}$.

(2) the half-angle formula $\sin A = \dfrac{2\tan\left(\dfrac{A}{2}\right)}{1 + \tan^2\left(\dfrac{A}{2}\right)}$.

In addition we shall use $\dfrac{\mathrm{d}}{\mathrm{d}\theta}\tan(a\theta + b) = a\sec^2(a\theta + b)$.

You may find it useful to verify this using differentiation by substitution — putting $x = a\theta + b$ in the standard case $\dfrac{\mathrm{d}}{\mathrm{d}x}(\tan x) = \sec^2 x$.

The first step in the solution is to simplify the denominator of the integral. If we compare this with the relationship in (1) above, then $a = 3$ and $b = 4$,

giving $R = \sqrt{3^2 + 4^2} = 5$ and $\tan \alpha = \frac{4}{3} = 1.333$

giving $\alpha = \arctan 1.333 = 0.927$ rad

It is simpler to use the symbol α throughout the integration, and then put in its numerical value when evaluating the integral in the last few lines of the solution.

Thus

$$\int_0^1 \frac{d\theta}{3 \sin \theta + 4 \cos \theta} = \int_0^1 \frac{d\theta}{5 \sin (\theta + \alpha)}$$

$$= \frac{1}{5} \int_0^1 \frac{1}{\sin (\theta + \alpha)} d\theta$$ where $t = \tan \left(\dfrac{\theta + \alpha}{2} \right)$

$$= \frac{1}{5} \int_{0.50}^{1.44} \frac{1}{\frac{2t}{(1+t^2)}} \times \frac{2\,dt}{(1+t^2)}$$ or $t = \tan \left(\dfrac{\theta}{2} + \dfrac{\alpha}{2} \right)$

Thus $\dfrac{dt}{d\theta} = \frac{1}{2}\sec^2 \left(\dfrac{\theta}{2} + \dfrac{\alpha}{2} \right)$

$$= \frac{1}{5} \int_{0.50}^{1.44} \frac{1}{t}\,dt$$

or $\dfrac{dt}{d\theta} = \frac{1}{2}\sec^2 \left(\dfrac{\theta + \alpha}{2} \right)$

$$= \frac{1}{5} \Big[\log_e t \Big]_{0.50}^{1.44}$$

$$= \frac{1}{5} (\log_e 1.44 - \log_e 0.50)$$ or $\dfrac{dt}{d\theta} = \frac{1}{2}\left\{ 1 + \tan^2 \left(\dfrac{\theta + \alpha}{2} \right) \right\}$

$$= \frac{1}{5} \left(\log_e \frac{1.44}{0.50} \right)$$ or $\dfrac{dt}{d\theta} = \frac{1}{2}(1 + t^2)$

$$= 0.212$$

$\therefore$ $d\theta = \dfrac{2\,dt}{(1 + t^2)}$

Change of limits:

When $\theta = 1$ $t = \tan \left(\dfrac{1 + 0.927}{2} \right)$

$= 1.44$

When $\theta = 0$ $t = \tan \left(\dfrac{0 + 0.927}{2} \right)$

$= 0.50$

EXERCISE

14·5

By using suitable trigonometrical substitutions find the following indefinite integrals and evaluate the definite integrals:

1 $\displaystyle\int \frac{1}{\sqrt{1-x^2}}\,dx$

2 $\displaystyle\int_0^2 \frac{dx}{1+x^2}$

3 $\displaystyle\int \frac{1}{4+x^2}\,dx$

4 $\displaystyle\int_4^5 \frac{6}{\sqrt{25-x^2}}\,dx$

5 $\displaystyle\int_0^{\sqrt{3}} \frac{dx}{9+x^2}$

6 $\displaystyle\int \frac{dx}{\sqrt{1-4x^2}}$ Hint: put $2x = \sin u$

7 $\displaystyle\int_1^2 \frac{dx}{1+9x^2}$

8 $\displaystyle\int_0^1 \frac{1}{4+3x^2}\,dx$

9 $\displaystyle\int_2^3 \frac{dx}{x^2\sqrt{9+x^2}}$

10 $\displaystyle\int \frac{d\theta}{\sin\theta + \cos\theta}$

11 $\displaystyle\int_{\pi/2}^{\pi} \frac{d\theta}{\sin\theta - 2\cos\theta}$

12 $\displaystyle\int_0^1 \sqrt{1-x^2}\,dx$

by putting $x = \sin u$ and then using a double-angle formula.

Integration Using Partial Fractions

This method depends on being able to express the given function in terms of its partial fractions. This was covered in Chapter 13. Normally finding these partial fractions would be the first step in the solution, but here, to avoid repetition, examples worked out previously have been used.

Example 14.20

Find $\displaystyle\int \frac{2x^2 + 7x - 17}{(x-1)(x-2)(x+3)}\,dx$.

Now from p. 177 we have

$$\frac{2x^2 + 7x - 17}{(x-1)(x-2)(x+3)} = \frac{2}{(x-1)} + \frac{1}{(x-2)} - \frac{1}{(x+3)}$$

Thus

$$\int \frac{2x^2 + 7x - 17}{(x-1)(x-2)(x+3)} \, dx = \int \frac{2}{(x-1)} \, dx + \int \frac{1}{(x-2)} \, dx - \int \frac{1}{(x+3)} \, dx$$

Now for the first integral

$$\int \frac{2}{(x-1)} \, dx = 2 \int \frac{1}{u} \, du \qquad \text{where} \quad u = x - 1$$

$$= 2 \log_e u + c \qquad \therefore \quad \frac{du}{dx} = 1$$

$$= 2 \log_e (x-1) + c \qquad \text{or} \quad du = dx$$

Similar substitutions may be used for the other integrals, giving

$$\int \frac{2x^2 + 7x - 17}{(x-1)(x-2)(x+3)} = 2 \log_e (x-1) + \log_e (x-2) - \log_e (x+3) + k$$

Note that only one constant of integration is necessary in the complete solution.

Example 14.21

Find $\displaystyle \int_2^3 \frac{6x^2 + 9x + 1}{(x+1)^2(x-1)} \, dx$.

Now from p. 178 we have

$$\int \frac{6x^2 + 9x + 1}{(x+1)^2(x-1)} = \frac{1}{(x+1)^2} + \frac{2}{(x+1)} + \frac{4}{(x-1)}$$

Thus

$$\int \frac{6x^2 + 9x + 1}{(x+1)^2(x-1)} \, dx = \int \frac{1}{(x+1)^2} \, dx + \int \frac{2}{(x+1)} \, dx + \int \frac{4}{(x-1)} \, dx$$

Now for the first integral

$$\int \frac{1}{(x+1)^2} \, dx = \int \frac{1}{t^2} \, dt \qquad \text{where} \quad t = x + 1$$

$$= \int t^{-2} \, dt \qquad \therefore \quad \frac{dt}{dx} = 1$$

$$\text{or} \quad dt = dx$$

$$= \frac{t^{-1}}{-1} + c$$

$$= -\frac{1}{t} + c$$

$$= -\frac{1}{(x+1)} + c$$

The remaining integrals need substitutions similar to those in Example 14.20, giving

$$\int_2^3 \frac{6x^2 + 9x + 1}{(x+1)^2(x-1)}\, dx = \left[\frac{-1}{(x+1)} + 2\log_e(x+1) + 4\log_e(x-1) \right]_2^3$$

$$= \left\{ \frac{-1}{(3+1)} + 2\log_e(3+1) + 4\log_e(3-1) \right\}$$

$$- \left\{ \frac{-1}{(2+1)} + 2\log_e(2+1) + 4\log_e(2-1) \right\}$$

$$= -\tfrac{1}{4} + 2\log_e 4 + 4\log_e 2 + \tfrac{1}{3} - 2\log_e 3 - 4\log_e 1$$

$$= 3.43$$

Example 14.22

Find $\int \dfrac{3x^2 + 5x + 1}{(x^2 + 4)(x+3)}\, dx$.

Now from pp. 179–80 we have

$$\frac{3x^2 + 5x + 1}{(x^2 + 4)(x+3)} = \frac{2x-1}{(x^2+4)} + \frac{1}{(x+3)}$$

Thus

$$\int \frac{3x^2 + 5x + 1}{(x^2 + 4)(x+3)}\, dx = \int \frac{2x-1}{(x^2+4)}\, dx + \int \frac{1}{(x+3)}\, dx$$

$$= \int \frac{2x}{(x^2+4)}\, dx - \int \frac{1}{(x^2+4)}\, dx + \int \frac{1}{(x+3)}\, dx$$

Now for the first integral

$$\int \frac{2x}{(x^2+4)}\,dx = \int \frac{2x}{u}\frac{du}{2x} \qquad\qquad \text{where} \quad u = x^2+4$$

$$= \int \frac{1}{u}\,du \qquad\qquad\qquad \therefore \quad \frac{du}{dx} = 2x$$

$$= \log_e u + c \qquad\qquad\qquad \text{or} \quad dx = \frac{du}{2x}$$

$$= \log_e (x^2+4) + c$$

Now for the second integral, which needs a trigonometrical substitution,

$$\int \frac{dx}{(x^2+4)} = \int \frac{2\sec^2\theta\,d\theta}{(2\tan\theta)^2+4} \qquad \text{where} \quad x = 2\tan\theta$$

$$= \int \frac{2\sec^2\theta\,d\theta}{4\tan^2\theta+4} \qquad\qquad \therefore \quad \frac{dx}{d\theta} = 2\sec^2\theta$$

$$= \int \frac{2\sec^2\theta\,d\theta}{4(1+\tan^2\theta)} \qquad\qquad \begin{array}{l}\text{or} \quad dx = 2\sec^2\theta\,d\theta \\ \text{and as} \quad x = 2\tan\theta\end{array}$$

$$= \int \frac{\sec^2\theta\,d\theta}{2(\sec^2\theta)} \qquad\qquad \text{then} \quad \theta = \arctan\frac{x}{2}$$

$$= \tfrac{1}{2}\int 1\,d\theta$$

$$= \tfrac{1}{2}\theta + k$$

$$= \tfrac{1}{2}\arctan\frac{x}{2} + k$$

The third integral is similar to those in Example 14.20, giving

$$\int \frac{3x^2+5x+1}{(x^2+4)(x+3)}\,dx = \log_e(x^2+4) - \tfrac{1}{2}\arctan\frac{x}{2} + \log_e(x+3) + c$$

EXERCISE

14·6

By the use of partial fractions find the following integrals and evaluate the definite integrals:

1 $\displaystyle\int \frac{5x-3}{(x-3)(x+3)}\,dx$

2 $\displaystyle\int_3^4 \frac{x}{(x-1)(x-2)}\,dx$

3 $\displaystyle\int \frac{9x^2+34x+29}{(x+1)(x+2)(x+3)}\,dx$

4 $\displaystyle\int_{1.1}^{1.5} \frac{3x^2-7x+2}{(2x-1)(x+1)(x-1)}\,dx$

5 $\int \dfrac{6 - 15x + 7x^2}{2x(1-x)(2-3x)}\,dx$ **6** $\int_2^5 \dfrac{2x+4}{(1+x)(x-1)(2x+1)}\,dx$

7 $\int \dfrac{7x - 2x^2}{(2-x)^2(1+x)}\,dx$ **8** $\int_0^2 \dfrac{x^2+3}{(x+1)(x-1)^2}\,dx$

9 $\int \dfrac{-(2+3x)}{x(1+x)^2}\,dx$ **10** $\int_0^3 \dfrac{3x^2 + 3x + 2}{(x^2+1)(x+1)}\,dx$

11 $\int_1^2 \dfrac{6x^2 + 8x + 16}{(x+2)(x^2+4)}\,dx$ **12** $\int_2^4 \dfrac{x^2 + 4x - 2}{(x-1)(x+2)}\,dx$

Integration by Parts

We often need to integrate the product of two functions, for example $\int xe^{2x}\,dx$, and so we shall find a rule to cover such cases.

Now for differentiating products we have

$$\text{if} \qquad y = u \times v$$

$$\text{then} \qquad \frac{dy}{dx} = v\frac{du}{dx} + u\frac{dv}{dx}$$

and if we integrate both sides with respect to x then

$$\int \frac{dy}{dx}\,dx = \int v\frac{du}{dx}\,dx + \int u\frac{dv}{dx}\,dx$$

or simplifying,

$$\int 1\,dy = \int v\,du + \int u\,dv$$

giving

$$y = \int v\,du + \int u\,dv$$

But $y = uv$, so

$$uv = \int v\,du + \int u\,dv$$

and rearranging

$$\boxed{\int u\,dv = uv - \int v\,du}$$

Example 14.23

Find $\int x e^{2x} \, dx$.

If we compare $\int x e^{2x} \, dx$ with the LHS of the rule $\int u \, dv$

we may put $u = x$ and $dv = e^{2x} \, dx$

Thus $\dfrac{du}{dx} = 1$ and $\int 1 \, dv = \int e^{2x} \, dx$

giving $du = dx$ and $v = \tfrac{1}{2} e^{2x}$

If we substitute these in the established rule

$$\int u \, dv = uv - \int v \, du$$

Then $\int x (e^{2x} \, dx) = x(\tfrac{1}{2} e^{2x}) - \int (\tfrac{1}{2} e^{2x}) \, dx$

$$= \tfrac{1}{2} x e^{2x} - \tfrac{1}{2} \int e^{2x} \, dx$$

Hence $\int x e^{2x} \, dx = \tfrac{1}{2} x e^{2x} - \tfrac{1}{4} e^{2x} + c$

As for the solution of all indefinite integrals we introduce one constant c after the final integration.

Beware! Suppose that the same integral had been given to you as $\int e^{2x} x \, dx$ and that, using the procedure followed previously,

comparing $\int e^{2x} x \, dx$ with the LHS of the rule $\int u \, dv$

gives $u = e^{2x}$ and $dv = x \, dx$

Then $\dfrac{du}{dx} = 2e^{2x}$ and $\int 1 \, dv = \int x \, dx$

giving $du = 2e^{2x} \, dx$ and $v = \dfrac{x^2}{2}$

and if we substitute these in the established rule

$$\int u \, dv = uv - \int v \, du$$

then $\int e^{2x}(x \, dx) = (e^{2x})\left(\dfrac{x^2}{2}\right) - \int \dfrac{x^2}{2} 2e^{2x} \, dx$

All is satisfactory except for the last term — this integral is not only a product but is more complicated than the original problem. So this choice for u and dv will not work. Thus if you had been given the problem as $\int e^{2x}x \, dx$ it would first have been necessary to change the order to $\int xe^{2x} \, dx$.

To avoid this disappointment two hints should be remembered:

(1) Always choose the 'u' function so that it will become simpler on differentiation.
(2) Always choose the 'dv' function that can be integrated easily.

You will not always be right first time but experience helps!

Example 14.24

Find $\displaystyle\int_{0.5}^{1} x^2 \log_e x \, dx$.

Comparing this with $\int u \, dv$ we could put $u = x^2$ and $dv = \log_e x \, dx$. Now the u term will become simpler on differentiation, but what about integrating the expression for dv? — horrible, to say the least. So we shall try writing the given product the other way round in the integral.

The problem is now $\displaystyle\int_{0.5}^{1} (\log_e x)x^2 \, dx$

It is easier to forget the limits until the integration has been completed and then reintroduce them.

Comparing this with $\int u \, dv$ we may put

$$u = \log_e x \quad \text{and} \quad dv = x^2 \, dx$$

thus $\dfrac{du}{dx} = \dfrac{1}{x}$ and $\int 1 \, dv = \int x^2 \, dx$

giving $du = \dfrac{1}{x} \, dx$ and $v = \dfrac{x^3}{3}$

Substituting these in the rule

$$\int u \, dv = uv - \int v \, du$$

Then $\int (\log_e x)(x^2 \, dx) = (\log_e x)\left(\dfrac{x^3}{3}\right) - \int \dfrac{x^3}{3}\left(\dfrac{1}{x} \, dx\right)$

$$= \tfrac{1}{3}x^3 \log_e x - \tfrac{1}{3}\int x^2 \, dx$$

$$= \tfrac{1}{3}x^3 \log_e x - \dfrac{x^3}{9} + c$$

Thus $\displaystyle\int_{0.5}^{1} x^2 \log_e x \, dx = \left[\dfrac{x^3}{3} \log_e x - \dfrac{x^3}{9}\right]_{0.5}^{1}$

$$= \left(\dfrac{1^3}{3} \log_e 1 - \dfrac{1^3}{9}\right) - \left(\dfrac{0.5^3}{3} \log_e 0.5 - \dfrac{0.5^3}{9}\right)$$

$$= -0.0683$$

Example 14.25

Find $\int \log_e x \, dx$.

We may use integration by parts by rewriting the integral as
$\int (\log_e x) \, 1 \, dx$.

Comparing this with $\int u \, dv$ we may put

$$u = \log_e x \quad \text{and} \quad dv = 1 \, dx$$

Thus $\dfrac{du}{dx} = \dfrac{1}{x} \quad \text{and} \quad \int 1 \, dv = \int 1 \, dx$

giving $du = \dfrac{1}{x} \, dx \quad \text{and} \quad v = x$

Substituting in $\int u \, dv = uv - \int v \, du$

gives $$\int (\log_e x)(1\ dx) = (\log_e x)x - \int x\left(\frac{1}{x}\ dx\right)$$

$$= (\log_e x)x - \int 1\ dx$$

$$= x\log_e x - x + c$$

$$= x(\log_e x - 1) + c$$

Example 14.26

Find $\int x^2 \sin 2x\ dx$.

Comparing this with $\int u\ dv$ we may put

$$u = x^2 \quad \text{and} \quad dv = \sin 2x\ dx$$

Thus $$\frac{du}{dx} = 2x \quad \text{and} \quad \int 1\ dv = \int \sin 2x\ dx$$

giving $$du = 2x\ dx \quad \text{and} \quad v = -\tfrac{1}{2}\cos 2x$$

Substituting in

$$\int u\ dv = uv - \int v\ du$$

gives $$\int (x^2)(\sin 2x\ dx) = x^2(-\tfrac{1}{2}\cos 2x) - \int (-\tfrac{1}{2}\cos 2x)(2x\ dx)$$

$$= -\frac{x^2}{2}\cos 2x + \int x\cos 2x\ dx$$

Now $\int x\cos 2x\ dx$ will need to be integrated by parts, giving

$$\int (x^2)(\sin 2x\ dx) = -\frac{x^2}{2}\cos 2x + \left\{x(\tfrac{1}{2}\sin 2x) - \int (\tfrac{1}{2}\sin 2x)\ dx\right\}$$

$$= -\frac{x^2}{2}\cos 2x + \frac{x}{2}\sin 2x + \tfrac{1}{4}\cos 2x + c$$

Example 14.27

Find $\int e^{3x} \cos x \, dx$.

Let us put $I = \int e^{3x} \cos x \, dx$. You will see how this proves useful later.

Comparing with $\int u \, dv$ we may put

$$u = e^{3x} \qquad \text{and} \qquad dv = \cos x \, dx$$

$$\text{Thus} \quad \frac{du}{dx} = 3e^{3x} \qquad \text{and} \qquad 1 \, dv = \cos x \, dx$$

$$\text{giving} \quad du = 3e^{3x} \, dx \quad \text{and} \qquad v = \sin x$$

Substituting in $\int u \, dv = uv - \int v \, du$

gives $\quad e^{3x}(\cos x \, dx) = e^{3x} \sin x - \int (\sin x)(3e^{3x} \, dx)$

or $\qquad\qquad I = e^{3x} \sin x - 3 \int e^{3x} \sin x \, dx$

Now $\int e^{3x} \sin x \, dx$ will need to be integrated by parts, giving

$$I = e^{3x} \sin x - 3\left\{ e^{3x}(-\cos x) - \int (-\cos x) 3e^{3x} \, dx \right\}$$

$$= e^{3x} \sin x + 3e^{3x} \cos x - 9 \int e^{3x} \cos x \, dx$$

$$\therefore \qquad I = e^{3x} \sin x + 3e^{3x} \cos x - 9I + c$$

Here we have an equation from which we wish to find I.

Thus $\qquad\qquad I + 9I = e^{3x} \sin x + 3e^{3x} \cos x + c$

or $\qquad\qquad 10I = e^{3x} \sin x + 3e^{3x} \cos x + c$

Thus $\qquad\qquad I = \tfrac{1}{10} e^{3x}(\sin x + 3 \cos x) + \dfrac{c}{10}$

and finally we may replace $\dfrac{c}{10}$, which is simply a constant, by k.

Hence $\qquad\qquad \int e^{3x} \cos x \, dx = \dfrac{e^{3x}}{10}(\sin x + 3 \cos x) + k$

EXERCISE

14·7

Using integration by parts, find the following indefinite integrals and evaluate the definite integrals:

1 $\int xe^x \, dx$

2 $\int 3x(\log_e 5x) \, dx$

3 $\int 2xe^{3x} \, dx$

4 $\int_0^{\pi/2} x \cos x \, dx$

5 $\int_{\pi/2}^{\pi} x \sin x \, dx$

6 $\int_1^2 \log_e x \, dx$

7 $\int_0^1 xe^{-x} \, dx$

8 $\int_0^{\pi} (\pi - x) \cos x \, dx$

9 $\int_1^2 \sqrt{x} \log_e x \, dx$

Using integration by parts twice, find the following integrals and evaluate the definite integrals:

10 $\int x^2 \cos x \, dx$

11 $\int x^2 \sin 3x \, dx$

12 $\int_0^{0.4} x^2 e^x \, dx$

13 $\int e^t \cos t \, dt$

14 $\int_0^1 e^{2t} \sin 3t \, dt$

15 $\int e^{ax} \cos bx \, dx$

Numerical Integration

OBJECTIVES

1 Derive the trapezium and mid-ordinate rules for numerical integration.
2 Derive Simpson's rule over two intervals.
3 Deduce the general form of Simpson's rule over an even number of equal intervals.
4 Evaluate definite integrals using the trapezium, mid-ordinate and Simpson's rules.
5 Determine answers to a desired accuracy using Simpson's rule.

Introduction

We know that the area under a curve between limits $x = a$ and $x = b$ is given by the sum of all the elementary strip areas (Fig. 15.1).

Fig. **15.1**

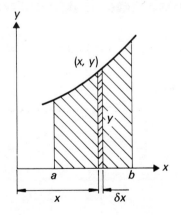

In mathematical notation

$$\text{Area} = \sum_{x=a}^{x=b} y\, \delta x \qquad \text{approximately}$$

or

$$\text{Area} = \int_a^b y\, dx \qquad \text{exactly}$$

217

However, it is not always possible to evaluate the integral by direct mathematical integration. For example, we may have a curve obtained from experimental results for which there is no equation giving y in terms of x. Another difficulty arises when the expression for y in terms of x is too complicated for us to integrate.

We may then have to resort to dividing the required area into a number of strips (similar to elementary strips), finding the area of each strip, and then adding these up. Therefore the result will depend on calculating, as accurately as possible, the areas of the vertical strips.

Three reasonably simple methods are available: trapezium, mid-ordinate and Simpson's rules. You may have used these previously, but we shall now examine, in more detail, their derivation and the accuracy of results obtained.

The worked examples, which appear later in this chapter, to illustrate the use of the three methods can also be solved by direct integration and thus exact results obtained. (Here 'exact' means 'as accurate as required' since the answers are decimal numbers.) You may find it useful to check these. This enables the error to be calculated for each result.

The Trapezium (or Trapezoidal) Rule

Consider the area having the boundary ABCD shown in Fig. 15.2.

Fig. **15.2**

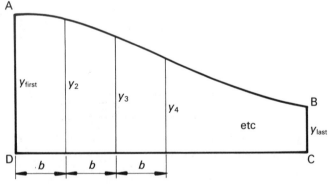

The area is divided into a number of vertical strips of equal width b.

Each vertical strip is assumed to be a trapezium. Hence the third strip, for example, will have an area $= b \times \frac{1}{2}(y_3 + y_4)$.

But

Area ABCD = The sum of all the vertical strips

$$= b \times \tfrac{1}{2}(y_{\text{first}} + y_2) + b \times \tfrac{1}{2}(y_2 + y_3) + b \times \tfrac{1}{2}(y_3 + y_4) + \dots$$

$$= b[\tfrac{1}{2}y_{\text{first}} + \tfrac{1}{2}y_2 + \tfrac{1}{2}y_2 + \tfrac{1}{2}y_3 + \dots + \tfrac{1}{2}y_{\text{last}}]$$

$$= b[\tfrac{1}{2}(y_{\text{first}} + y_{\text{last}}) + y_2 + y_3 + y_4 \dots]$$

> Area ABCD = Strip width × [$\tfrac{1}{2}$(Sum of first and last ordinates)
> + (Sum of remaining ordinates)]

This is known as the **trapezium rule**.

Example 15.1

Find $\displaystyle\int_0^{1.5} e^x \, dx$ by numerical integration using the trapezium rule.

The integral represents the area under the curve $y = e^x$ as shown in Fig. 15.3. The area has been divided into six vertical strips of equal width and the lengths of the ordinates found using a calculator.

Fig. **15.3**

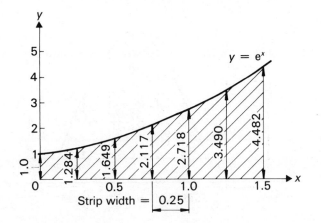

Now the trapezium rules states

$$\text{Area} = (\text{Strip width}) \times [\tfrac{1}{2}(\text{Sum of first and last ordinates}) + (\text{Sum of remaining ordinates})]$$

Thus

$$\int_0^{1.5} e^x \, dx = \text{Area under the curve}$$

$$\approx 0.25 \times [\tfrac{1}{2}(1.000 + 4.482) + (1.284 + 1.649 + 2.117 + 2.718 + 3.490)]$$

$$= 3.500 \text{ correct to 3 decimal places}$$

The result by integration is 3.482 correct to 3 decimal places.

∴   Error $= \dfrac{3.500 - 3.482}{3.482} \times 100 = +0.52\%$

Example 15.2

Find $\displaystyle\int_0^{\pi} \sin \theta \, d\theta$ by numerical integration using the trapezium rule.

The integral represents the area under the curve $y = \sin \theta$ as shown in Fig. 15.4. The area has been divided into six vertical strips of equal width and the lengths of the ordinates found using a calculator.

Fig. **15.4**

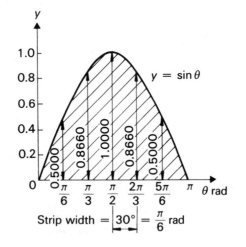

Now the trapezium rule states

Area $= (\text{Strip width}) \times [\tfrac{1}{2}(\text{Sum of first and last ordinates})$
$\qquad\qquad\qquad\qquad + (\text{Sum of remaining ordinates})]$

Thus

$$\int_0^\pi \sin\theta \, d\theta = \text{Area under the curve}$$

$$\approx \frac{\pi}{6} \times [\tfrac{1}{2}(0+0) + (0.5000 + 0.8660 + 1.0000 + 0.8660$$
$$+ 0.5000)]$$

$$= 1.954 \quad \text{correct to 3 decimal places}$$

The result by integration is 2.000 correct to 3 decimal places.

$$\therefore \qquad\qquad \text{Error} = \frac{1.954 - 2.000}{2.000} \times 100 = -2.3\%$$

The Mid-ordinate Rule

Consider the area having the boundary ABCD shown in Fig. 15.5.

Fig. **15.5**

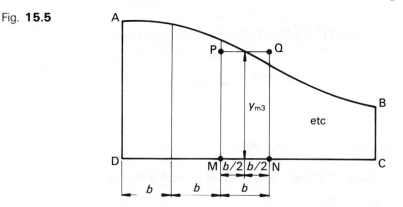

The area is divided into a number of vertical strips of equal width b.

In the third strip the vertical line of length y_{m3} is half way between the boundary ordinates and is called the mid-ordinate of the third strip.

The area of the third strip is assumed to be the same as the area of the rectangle PQNM which is (MN × PM) or by_{m3}.

Thus, if the mid-ordinates of all the strips are y_{m1}, y_{m2}, y_{m3}, etc.

then $\qquad\qquad$ Area ABCD $= by_{m1} + by_{m2} + by_{m3} + ...$

$$= b(y_{m1} + y_{m2} + y_{m3} + ... + y_{m\,\text{last}})$$

or $\qquad$ Area ABCD = (Strip width) × (Sum of mid-ordinates)

Example 15.3

Find $\displaystyle\int_0^{1.5} e^x\,dx$ by numerical integration using the mid-ordinate rule.

The integral represents the area under the curve $y = e^x$ as shown in Fig. 15.6. The area has been divided into six vertical strips of equal width and the lengths of the mid-ordinates found using a calculator.

Fig. **15.6**

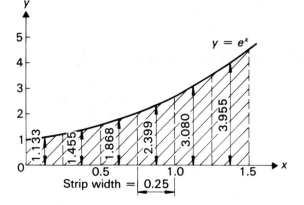

Now the mid-ordinate rule states

$$\text{Area} = (\text{Strip width}) \times (\text{Sum of mid-ordinates})$$

Thus

$$\int_0^{1.5} e^x\,dx = \text{Area under the curve}$$

$$\approx 0.25 \times (1.133 + 1.455 + 1.868 + 2.399 + 3.080 + 3.955)$$

$$= 3.473 \quad \text{correct to 3 decimal places}$$

The result by integration is 3.482 correct to 3 decimal places.

$$\therefore \qquad \text{Error} = \frac{3.473 - 3.482}{3.482} \times 100 = -0.26\%$$

Example 15.4

Find $\displaystyle\int_0^\pi \sin\theta\,d\theta$ by numerical integration using the mid-ordinate rule.

The integral represents the area under the curve $y = \sin\theta$ as shown in Fig. 15.7. The area has been divided into six vertical strips of equal⋅ width and the lengths of the mid-ordinates found using a calculator.

Fig. **15.7**

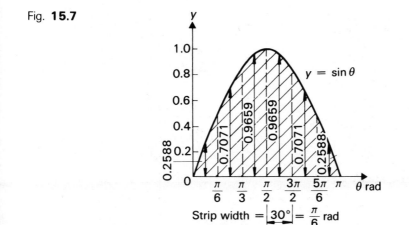

Now the mid-ordinate rule states

$$\text{Area} = (\text{Strip width}) \times (\text{Sum of mid-ordinates})$$

Thus

$$\int_0^\pi \sin\theta\,d\theta = \text{Area under the curve}$$

$$\approx \frac{\pi}{6} \times (0.2588 + 0.7071 + 0.9659 + 0.9659 + 0.7071 + 0.2588)$$

$$= 2.023 \quad \text{correct to 3 decimal places}$$

The result by integration is 2.000 correct to 3 decimal places.

$$\therefore \quad \text{Error} = \frac{2.023 - 2.000}{2.000} \times 100 = +1.15\%$$

Simpson's Rule

Suppose we wish to find the area PQRNL shown shaded in Fig. 15.8. QM is the mid-ordinate which divides the area into two strips of equal width, b. Fig. 15.9 is a repeat of Fig. 15.8 but shows a straight line AB drawn through point Q and parallel to chord PR.

Fig. **15.8**   Fig. **15.9**

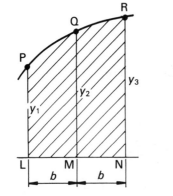
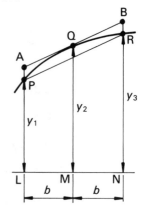

Now

$$\left(\begin{array}{c}\text{Area of}\\\text{parallelogram ABRP}\end{array}\right) = \left(\begin{array}{c}\text{Area of}\\\text{trapezium ABNL}\end{array}\right) - \left(\begin{array}{c}\text{Area of}\\\text{trapezium PRNL}\end{array}\right)$$

$$= 2by_2 - \tfrac{1}{2}(y_1 + y_3)2b$$

$$= b(2y_2 - y_1 - y_3)$$

Now the area PQR, between the curve and the chord, is taken as $\frac{2}{3}$ of the area of the parallelogram ABRP. This is a very close approximation. (It is only exact if the curve is the arc of a parabola and also AB is the tangent at Q.)

Thus $\text{Area PQRNL} = \left(\begin{array}{c}\text{Area of}\\\text{trapezium PRNL}\end{array}\right) + \text{Area PQR}$

$$= \tfrac{1}{2}(y_1 + y_3)2b + \tfrac{2}{3}b(2y_2 - y_1 - y_3)$$

$$= b(y_1 + y_3 + \tfrac{4}{3}y_2 - \tfrac{2}{3}y_1 - \tfrac{2}{3}y_3)$$

$$= \frac{b}{3}(y_1 + 4y_2 + y_3)$$

Fig. 15.10 shows a larger area divided into six vertical strip areas of equal width. Each adjacent pair of strips forms an area similar to that shown in Fig. 15.8. We therefore apply the expression we derived for each pair of strips as follows:

$$\text{Shaded area} = \begin{pmatrix} \text{Area between} \\ \text{ordinates} \\ y_1 \text{ and } y_3 \end{pmatrix} + \begin{pmatrix} \text{Area between} \\ \text{ordinates} \\ y_3 \text{ and } y_5 \end{pmatrix} + \begin{pmatrix} \text{Area between} \\ \text{ordinates} \\ y_5 \text{ and } y_7 \end{pmatrix}$$

$$= \frac{b}{3}(y_1 + 4y_2 + y_3) + \frac{b}{3}(y_3 + 4y_4 + y_5) + \frac{b}{3}(y_5 + 4y_6 + y_7)$$

$$= \tfrac{1}{3}b[y_1 + y_7 + 4(y_2 + y_4 + y_6) + 2(y_3 + y_5)]$$

Fig. **15.10**

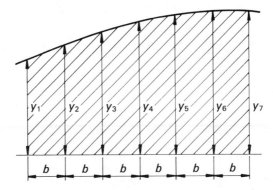

This idea may be extended providing there is an *even* number of strips of equal width to give a general expression for Simpson's rule:

> Area = $\frac{1}{3}$(Strip width) × [(Sum of first and last ordinates)
> + 4(Sum of even ordinates)
> + 2(Sum of remaining odd ordinates)]

Example 15.5

Find $\displaystyle\int_0^{1.5} e^x \, dx$ by numerical integration using Simpson's rule.

In order to use Simpson's rule the required area must be divided into an *even* number of strips. This has been done in Fig. 15.3 and we will use the values of the ordinates given.

Now Simpson's rule states

$$\text{Area} = \tfrac{1}{3}(\text{Strip width}) \times [(\text{Sum of first and last ordinates})$$
$$+ 4(\text{Sum of even ordinates})$$
$$+ 2(\text{Sum of remaining odd ordinates})]$$

Thus

$$\int_0^{1.5} e^x \, dx = \text{Area under the curve}$$

$$\approx \tfrac{1}{3} \times 0.25 \times [(1.000 + 4.482)$$
$$+ 4(1.284 + 2.117 + 3.490)$$
$$+ 2(1.649 + 2.718)]$$

$$= 3.482 \quad \text{correct to 3 decimal places}$$

The result by integration is 3.482 correct to 3 decimal places and so there is no error.

An alternative layout using a table to show the calculations is often used.

Ordinate number	Length	Simpson's multiplier	Product
1	1.000	1	1.000
2	1.284	4	5.136
3	1.649	2	3.298
4	2.117	4	8.468
5	2.718	2	5.436
6	3.490	4	13.960
7	4.482	1	4.482
		Total product =	41.780

Hence $\text{Result} = \tfrac{1}{2} \times 0.25 \times 41.780 = 3.482$

Example 15.6

Find $\displaystyle\int_0^{\pi} \sin \theta \, d\theta$ by numerical integration using Simpson's rule.

In order to use Simpson's rule the required area must be divided into an *even* number of strips. This has been done in Fig. 15.4 and we will use the values of the ordinates given.

Now Simpson's rule states

Area = $\frac{1}{3}$(Strip width) × [(Sum of first and last ordinates)
+ 4(Sum of even ordinates)
+ 2(Sum of remaining odd ordinates

Thus

$$\int_0^\pi \sin\theta \, d\theta = \text{Area under the curve}$$

$$\approx \frac{1}{3} \times \frac{\pi}{6} \times [(0+0) + 4(0.5000 + 1.0000 + 0.5000) + 2(0.8660 + 0.8660)]$$

$$= 2.001 \quad \text{correct to 3 decimal places}$$

The result by integration is 2.000 correct to 3 decimal places.

$$\therefore \quad \text{Error} = \frac{2.001 - 2.000}{2.000} \times 100 = 0.05\%$$

Comparison of Trapezium, Mid-ordinate and Simpson's Rules

If a curve is shaped ⌢ then the trapezium rule gives the shaded area shown in Fig. 15.11. This area is smaller than the true area under the curve, and this is confirmed by the answer to Example 15.2 which is 2.3% too small.

Fig. **15.11**   Fig. **15.12**

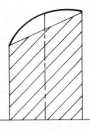
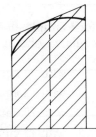

The mid-ordinate rule gives the shaded area shown in Fig. 15.12, which is greater than the true area under the curve. This is confirmed by the answer to Example 15.4 which is 1.15% too large.

For a curve shaped ⌣ the trapezium rule gives results too large (see Example 15.1), whilst the mid-ordinate rule gives results too small (see Example 15.3).

Simpson's rule, which is a combination of both the other rules, is much more accurate, as confirmed by the results of Examples 15.5 and 15.6. Hence if accuracy is of importance then Simpson's rule should be used.

Accuracy of Answers Using Simpson's Rule

In the preceding text it was possible to obtain exact answers to the worked examples. This enabled us to compare results obtained using approximate methods so that we may gain some idea of the degree of accuracy which may be expected.

However, in practice, we would only use Simpson's rule for integration if it were *not* possible to get an exact answer.

The accuracy of a result using Simpson's rule will depend on the number of intervals chosen. There are expressions which may be used to estimate the error, and also to decide on the number of intervals required to obtain an answer to a particular degree of accuracy. However, these are tedious to use and it is recommended that the following method be used:

> If an increase in the number of equal intervals does not involve a change in the answer to a certain degree of accuracy, then we may rely on the result to that degree of accuracy.

Suppose, for example, that for a particular problem when using Simpson's rule we obtain (correct to 3 significant figures):

with   6 equal intervals an answer of 6.36

with   8 equal intervals an answer of 6.34

with 10 equal intervals an answer of 6.33

and   with 12 equal intervals an answer of 6.33

Here we may safely assume the result is 6.33 correct to 3 significant figures.

EXERCISE

15·1

Evaluate, stating the answers to 3 significant figures, the following integrals, using:

(a) the trapezium rule
(b) the mid-ordinate rule
(c) Simpson's rule.

1 $\int_0^3 \sqrt{(9-x^2)}\, dx$ using 6 intervals

2 $\int_0^{1.5} \dfrac{dx}{\sqrt{(4-x^2)}}$ using 6 intervals

3 $\int_1^2 (\log_e x)\, dx$ using 10 intervals

4 $\int_0^{\pi/4} \left(\dfrac{\theta}{\cos\theta}\right) d\theta$ using 6 intervals

5 $\int_{0.1}^{0.7} \left(\dfrac{t}{1-t}\right) dt$ using 6 intervals

6 $\int_0^{\pi/6} (\sin^3 \phi)\, d\phi$ using 6 intervals

7 $\int_0^2 (x^2 e^{-x})\, dx$ using 8 intervals

8 Use Simpson's rule to find the value of

$$\int_{1.8}^3 \left(\dfrac{1}{\sin x}\right) dx$$

giving the answer correct to 3 significant figures. Start with 6 equal intervals and proceed until the required accuracy is guaranteed.

Differential Equations

OBJECTIVES

1   Determine and sketch a family of curves, given their derivative, for a simple function.

2   Determine a particular curve of the family by specifying a point on it.

3   Define a boundary condition.

4   Solve differential equations of the type $\dfrac{dy}{dx} = f(x)$ given a boundary condition.

5   Differentiate $y = Ae^{kx}$.

6   Verify that $y = Ae^{kx}$ satisfies $\dfrac{dy}{dx} = ky$ by substitution.

7   Derive equations of the form $\dfrac{dy}{dx} = ky$ from problems arising in technology.

8   Solve the derived equations in **7** using boundary conditions.

9   Define the terms differential equation (DE), order and degree.

10 Formulate DEs by eliminating constants A and B from an equation of the type $f(x, y, A, B) = 0$.

11 Formulate DEs from geometrical and physical problems, e.g. particle falling under gravity, chemical reaction, voltage in a circuit, etc.

12 Solve DEs of the form $\dfrac{dy}{dx} = f(x)$ and $\dfrac{dy}{dx} = f(y)$ by direct integration.

13 Define the term separable variables for a DE of the form $\dfrac{dy}{dx} = f(x), g(y)$.

14 Solving equations of the type in **13** by separating the variables and integrating to find the general solution.

15 Apply boundary conditions to the general solution to obtain a particular solution.

Families of Curves

Suppose we know that $\qquad \dfrac{dy}{dx} = 3$

This may be rewritten as $\qquad y = \displaystyle\int 3 \, dx$

from which $\qquad y = 3x + C$

The constant of integration C represents any number.

Now suppose we give C different values and plot the graph of each equation (Fig. 16.1).

Fig. **16.1**

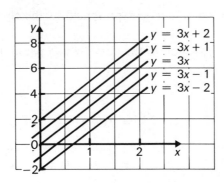

When $\qquad\qquad\qquad\quad C = -2:\qquad y = 3x - 2$

and when $\qquad\qquad\quad C = -1:\qquad y = 3x - 1$

and when $\qquad\qquad\quad C = 0:\qquad y = 3x$

and when $\qquad\qquad\quad C = +1:\qquad y = 3x + 1$

and when $\qquad\qquad\quad C = +2:\qquad y = 3x + 2$

It follows that the equation $y = 3x + C$ represents a set of graphs called a **family**. They all have one thing in common, that is $\dfrac{dy}{dx} = 3$.

If we specify a particular point we may find the equation of the graph which passes through that point.

Example 16.1

Find the equation of the graph which is one of the family represented by the equation $y = 3x + C$ and which passes through the point $(1, 7)$.

If a point lies on a graph its coordinates satisfy the equation of the graph.

Hence since the point $(1, 7)$ lies on the graph $y = 3x + C$

here $\qquad\qquad\qquad\qquad 7 = 3(1) + C$

from which $\qquad\qquad\qquad C = 4$

and therefore the required equation is $y = 3x + 4$.

Example 16.2

Sketch the family of curves represented by the equation $\dfrac{dy}{dx} = 2x + 1$.

Find also the equation of the curve which passes through the point (3, 18).

We have $$\frac{dy}{dx} = 2x + 1$$

or $$y = \int (2x + 1)\, dx$$

from which $$y = x^2 + x + C$$

This is the general solution of the given equation and represents the family of curves of which five typical ones are shown in Fig. 16.2.

Fig. **16.2**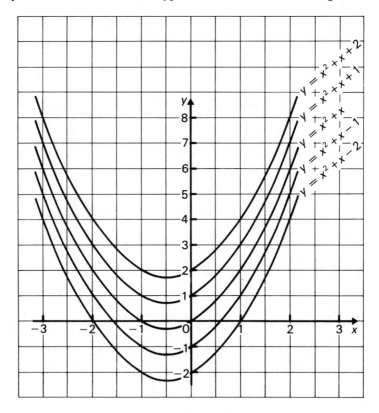

To find the equation of the curve passing through the point (3, 18) we must substitute these values of x and y in the general solution.

Hence $\qquad\qquad 18 = 3^2 + 3 + C$

$\therefore \qquad\qquad C = 6$

and therefore the required equation is $y = x^2 + x + 6$.

Basic Differential Equations

Equations which contain a differential coefficient such as

$$\frac{dy}{dx} = 3 \quad \text{or} \quad \frac{dy}{dx} = 2x + 1$$

are called **differential equations**.

The expressions obtained by integration, for y in terms of x, and which include unknown constants, are called **general solutions**.

If a value of y corresponding to a value of x is known then this is called a **boundary condition**. Values given by a boundary condition may be substituted into a general solution and a numerical value obtained for the constant. The resulting equations, such as

$$y = 3x + 4 \quad \text{and} \quad y = x^2 + x + 6$$

are called **particular solutions**.

Example 16.3

(a) Find the general solution of the differential equation

$$\frac{dy}{dx} = 5x^3 + 2x - 7$$

(b) From the boundary condition $y = 2$ when $x = 1$, find the particular solution.

(a) We have $\qquad\qquad \dfrac{dy}{dx} = 5x^3 + 2x - 7$

which may be rewritten as

$$y = \int (5x^3 + 2x - 7)\, dx$$

$$\therefore \qquad\qquad y = \tfrac{5}{4}x^4 + x^2 - 7x + C$$

This is the required general solution.

(b) To find the particular solution we must substitute $y = 2$ when $x = 1$ into the general solution.

$$\therefore \qquad 2 = \tfrac{5}{4}(1)^4 + (1)^2 - 7(1) + C$$

from which $\qquad C = 2 - \tfrac{5}{4} - 1 + 7$

$$= 6.75$$

Hence the required particular solution is

$$y = \tfrac{5}{4}x^4 + x^2 - 7x + 6.75$$

EXERCISE

16·1

1 Sketch the family of curves represented by the differential equation $\dfrac{dy}{dx} = x$, and find the equation of the curve passing through the point $(2, 3)$.

2 Sketch the family of curves represented by the differential equation $\dfrac{dy}{dx} = 3x^2$. Find also the equation of the curve which passes through the point $(5, -3)$.

3 Find the general solution of the differential equation

$$\frac{dy}{dx} = x^2 - 5x$$

and the particular solution if $x = 4$ when $y = 0$.

4 If $\dfrac{dy}{dx} = 6x^3 + 5x^2 + 7$, find the particular solution if it represents the equation of the curve which passes through the point $(2, 2)$.

5 If $\dfrac{dy}{dx} = ax$, where a is a constant, find the particular solution if $x = 0$ and $y = 0$, and also $x = 2$ when $y = 4$.

Equations of the Type $\dfrac{dy}{dx} = ky$

We will show now that $y = Ae^{kx}$ is the general solution of the differential equation $\dfrac{dy}{dx} = ky$. In order to do this we must prove that $y = Ae^{kx}$ satisfies the equation $\dfrac{dy}{dx} = ky$.

Now $\qquad\qquad\qquad y = Ae^{kx}$ [1]

and differentiating with respect to x we have

$$\frac{dy}{dx} = Ake^{kx}$$ [2]

Now for the given differential equation

$$\begin{aligned} \text{LHS} &= \frac{dy}{dx} \\ &= Ake^{kx} \quad \text{from Equation [2]} \\ &= k(Ae^{kx}) \\ &= ky \qquad \text{from Equation [1]} \\ &= \text{RHS} \end{aligned}$$

Hence the differential equation $\dfrac{dy}{dx} = ky$

has a general solution $\qquad y = Ae^{kx}$ where A is a constant.

Example 16.4

Given the differential equation $\dfrac{dy}{dx} = 6y$ find:

(a) the general solution
(b) the particular solution if $x = 0$ when $y = 3$

(a) Here the equation is of the form $\dfrac{dy}{dx} = ky$ where $k = 6$. Hence the general solution is $y = Ae^{6x}$.

(b) Substituting the values $x = 0$ and $y = 3$ into the general solution we have

$$3 = Ae^{6 \times 0}$$

from which $\qquad A = 3 \quad$ since $\quad e^0 = 1$

Hence the particular solution is $y = 3e^{6x}$.

Example 16.5

In a particular problem the rate of change of distance s with respect to time t is known to be proportional to the distance s. Express this as a differential equation and find:

(a) the particular solution if $s = 2$ when $t = 1$, and $s = 6$ when $t = 1.5$

(b) the value of s when $t = 3$

In mathematical notation the rate of change of s with respect to t is $\dfrac{ds}{dt}$.

Now we are given that $\dfrac{ds}{dt}$ is proportional to s,

that is, $\qquad\qquad \dfrac{ds}{dt} = ks, \quad$ where k is a constant.

The general solution of this differential equation is $s = Ae^{kt}$ where both A and k are constants. The boundary conditions given in part (a) will enable us to find the values of these two constants.

(a) Now $s = 2$ when $t = 1$, and substituting these values into the general solution we have

$$2 = Ae^{k \times 1} \qquad\qquad [1]$$

Similarly, since $s = 6$ when $t = 1.5$,

$$6 = Ae^{k \times 1.5} \qquad\qquad [2]$$

Dividing Equation [2] by Equation [1] gives

$$\frac{6}{2} = \frac{e^{1.5k}}{e^k}$$

$\therefore \qquad\qquad 3 = e^{(1.5-1)k}$

or $\qquad\qquad 3 = e^{0.5k}$

To solve this equation for k we must put it into logarithmic form,

that is $0.5k = \log_e 3$

∴ $k = \dfrac{1.10}{0.5}$

∴ $k = 2.20$

To find A we substitute $k = 2.20$ into Equation [1].

∴ $2 = Ae^{2.20}$

or $A = \dfrac{2}{9.03}$

∴ $A = 0.221$

Hence the required solution is $s = 0.221e^{2.20t}$

(b) When $t = 3$ the corresponding value of s may be found by substituting $t = 3$ into the particular solution:

$$s = 0.221e^{2.20 \times 3}$$

$$= 162$$

Example 16.6

Figure 16.3 shows the general arrangement of a belt drive.

Fig. **16.3**

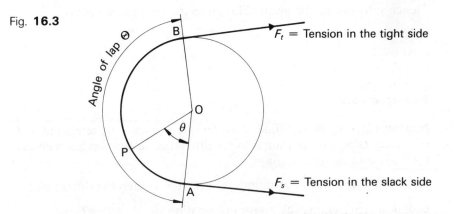

It can be shown that the rate of change of the tension F N with respect to the angle θ radians at any point P is given by μF, where μ is the coefficient of friction between the belt and the pulley.

Find the equation connecting F_t, F_s, μ and Θ and hence find the value of F_t when $F_s = 50$ N, $\mu = 0.3$ and the angle of lap is 120°.

The initial differential equation will be

$$\frac{\mathrm{d}F}{\mathrm{d}\theta} = \mu F$$

and hence the general solution will be of the form

$$F = A\mathrm{e}^{\mu\theta}$$

Now, when $\theta = 0$, $F = F_s$

∴ $F_s = A\mathrm{e}^{\mu \times 0}$

and since $\mathrm{e}^0 = 1$

$$A = F_s$$

and hence the solution becomes

$$F = F_s\mathrm{e}^{\mu\theta}$$

Also, when $\theta = \Theta$, $F = F_t$

∴ $F_t = F_s\mathrm{e}^{\mu\Theta}$

Hence the required equation is $F_t = F_s\mathrm{e}^{\mu\Theta}$

As in many mathematical equations the angle must be expressed in radians.

Since $360° = 2\pi$ radians

then $120° = \frac{2}{3}\pi = 2.09$ radians

Hence substituting the given values into the solution we have

$$F_t = 50\mathrm{e}^{0.3 \times 2.09}$$

$$= 93.6 \text{ N}$$

Example 16.7

Newton's law of cooling states that the rate at which the temperature T of a body falls is proportional to the difference in temperature between the body and its surroundings.

If t is the time and the surroundings are at $0\,°C$ then the differential equation representing the above information is $\dfrac{\mathrm{d}T}{\mathrm{d}t} = -kT$, the negative sign indicating a temperature drop.

If a body's temperature falls from $90\,°C$ to $70\,°C$ in 50 seconds find how long it will take to cool another $20\,°C$.

Since the differential equation given is

$$\frac{dT}{dt} = -kT$$

the general solution is of the form

$$T = Ae^{-kt}$$

If we assume that cooling starts at 90 °C, then $T = 90$ when $t = 0$ s, and substituting these values into the general equation we have:

$$90 = Ae^{-k \times 0}$$

and since $e^0 = 1$, $A = 90$

Also, when $T = 70$ °C then $t = 50$ s, and substituting these values into $T = 90e^{-kt}$ we get

$$70 = 90e^{-k \times 50}$$

$$\therefore \qquad e^{50k} = \frac{90}{70}$$

and rearranging into logarithmic form gives us

$$50k = \log_e\left(\frac{90}{70}\right)$$

$$\therefore \qquad k = 0.005\ 03$$

Hence the particular solution is

$$T = 90e^{-0.005\ 03t}$$

We shall now find the time to cool another 20 °C, that is to 50 °C. If we substitute $T = 50$ °C into $T = 90e^{-0.005\ 03t}$ the value of t will be the time to cool from 90 °C, that is when the time commences, to 50 °C.

Substituting gives us $50 = 90e^{-0.005\ 03t}$

$$e^{-0.005\ 03t} = \frac{90}{50}$$

and rearranging into logarithmic form

$$0.005\ 03t = \log_e\left(\frac{90}{50}\right)$$

$$t = 117\ s$$

Hence the time to cool from 70 °C to 50 °C $= 117 - 50$

$$= 67 \text{ seconds}$$

EXERCISE

16·2

1 Find the general solution of the differential equation
$\dfrac{dy}{dx} = 3y$ and the particular solution if $x = 0$ when $y = 2$.
Find also the value of y when $x = 2$.

2 Find the particular solution of the differential equation
$\dfrac{ds}{dt} = -ks$ if $s = 7$ when $t = 1$, and also $s = 4$ when
$t = 2$. Find also the value of t when $s = 2$.

3 The rate at which the current I A dies in an electrical circuit
which contains a resistance $R\,\Omega$ and an inductance L H is
given by the differential equation $\dfrac{dI}{dt} = -\dfrac{R}{L}I$ where t s is
time. If $R = 2\,\Omega$ and $L = 0.06$ H, and $I = 10$ A when
$t = 0$ s, find the solution of the differential equation. Find
also the current after 0.02 s.

4 A radioactive material decays at a rate which is proportional
to the amount of radioactivity remaining. If the amount of
radioactivity remaining is denoted by N at the time t, form a
differential equation which represents this statement. If the
half-life (that is the time taken for half of the radioactivity
to decay) is 10 years, how long will it take for $\frac{3}{4}$ of the
radioactivity to decay?

5 An electrical circuit contains a resistance R ohms and a
capacitor C farads which initially holds a charge Q coulombs.
The rate of discharge of the capacitor is given by the
equation $\dfrac{dQ}{dt} + \dfrac{Q}{RC} = 0$, where t seconds is the time. If
$R = 80\,000$ ohms and $C = 0.3 \times 10^{-6}$ farads and also
$Q = 0.0015$ coulombs initially, find the equation connecting
Q and t. Find Q when $t = 0.02$ seconds and also t when
$Q = 0.001$ coulombs.

Differential Equations — Their Order and Degree

Previously we have met equations which contain differential coefficients, or derivatives, such as $\dfrac{\mathrm{d}y}{\mathrm{d}x} = x + 6$. These are called **differential equations**. For brevity 'differential equation' is simply stated as 'DE'.

Now $\dfrac{\mathrm{d}y}{\mathrm{d}x}$ is known as a first derivative

and $\dfrac{\mathrm{d}^2 y}{\mathrm{d}x^2}$ is known as a second derivative

and $\dfrac{\mathrm{d}^3 y}{\mathrm{d}x^3}$ is known as a third derivative, and so on.

Now DEs are classified according to their order. The order of a DE is given by the highest derivative that occurs in the equation.

Thus $\dfrac{\mathrm{d}y}{\mathrm{d}x} = x + 6$ is a DE of the first order,

$\dfrac{\mathrm{d}^2 y}{\mathrm{d}x^2} = e^{ax}$ is a DE of the second order,

$\dfrac{\mathrm{d}^2 y}{\mathrm{d}x^2} + 6\dfrac{\mathrm{d}y}{\mathrm{d}x} - y = 0$ is a DE of the second order,

$\dfrac{\mathrm{d}^3 y}{\mathrm{d}x^3} + x\dfrac{\mathrm{d}y}{\mathrm{d}x} = 7y$ is a DE of the third order.

The degree of a DE is the degree of the highest derivative.

Hence $\left(\dfrac{\mathrm{d}^2 y}{\mathrm{d}x^2}\right)^2 + \dfrac{\mathrm{d}y}{\mathrm{d}x} + 3y = 0$ is a DE of the second order and second degree,

$\dfrac{\mathrm{d}^3 y}{\mathrm{d}x^2} = \left(\dfrac{\mathrm{d}y}{\mathrm{d}x}\right)^2$ is a DE of the third order and first degree,

$\left(\dfrac{\mathrm{d}^4 y}{\mathrm{d}x^4}\right)^3 = x$ is a DE of the fourth order and third degree.

This all sounds very frightening but you will soon become familiar with the expressions and terms used.

EXERCISE

16·3

State the order and degree of the following DEs.

1 $\dfrac{dy}{dx} = y^2$

2 $\dfrac{d^3y}{dx^3} + \dfrac{d^2y}{dx^2} = 0$

3 $\dfrac{d^2y}{dx^2} - \left(\dfrac{dy}{dx}\right)^3 = 6x$

4 $\left(\dfrac{dy}{dx}\right)^2 + 2y = 3$

5 $\dfrac{d^4y}{dx^4} + \dfrac{d^2y}{dx^2} = \sin 2x$

6 $\dfrac{d^2y}{dx^2} = \left(\dfrac{dy}{dx}\right)^2$

7 $\dfrac{d^2y}{dx^2} - 6\dfrac{dy}{dx} + y = 0$

8 $\left(\dfrac{d^3y}{dx^3}\right)^2 + \left(\dfrac{dy}{dx}\right)^3 + y^4 = 0$

9 $\dfrac{d^4y}{dx^4} - 2\dfrac{d^3y}{dx^3} - \dfrac{dy}{dx} = e^x$

10 $\dfrac{d^2y}{dx^2} = \left(\dfrac{dy}{dx}\right)^3$

Differential Equations Formed by Eliminating Constants

The standard equation of a straight line, where m (the gradient) and c are constants, is given by

$$y = mx + c$$

Now

$$\frac{dy}{dx} = m$$

and

$$\frac{d^2y}{dx^2} = 0$$

Now $\dfrac{dy}{dx} = m$ is a first order DE which indicates that, at any point on the line, the gradient is constant and equal to m.

Similarly $\dfrac{d^2y}{dx^2} = 0$ is a second order DE which indicates there is a zero rate of change of gradient with respect to x. In other words the gradient remains unaltered as m.

We know that the statements are true but it is interesting to see how they are represented by DEs. It should also be noted that each time we differentiate a constant disappears, first c and then m.

Example 16.8

The laws of growth of bacteria or compound interest are based on an equation of the type $y = ae^{bt}$ where a and b are constants. By eliminating the constants find a differential equation which expresses this relationship.

Starting with $$y = ae^{bt} \qquad [1]$$

then $$\frac{dy}{dt} = abe^{bt} \qquad [2]$$

and $$\frac{d^2y}{dt^2} = ab^2e^{bt} \qquad [3]$$

Equation [2] may be rewritten as $$\frac{dy}{dt} = b(ae^{bt}) = by$$

from which $$b = \frac{1}{y}\left(\frac{dy}{dt}\right) \qquad [4]$$

Equation [3] may be rewritten as $$\frac{d^2y}{dt^2} = b(abe^{bt})$$

∴ using Equations [4] and [2] $$\frac{d^2y}{dt^2} = \frac{1}{y}\left(\frac{dy}{dt}\right) \times \left(\frac{dy}{dt}\right)$$

from which $$y\left(\frac{d^2y}{dt^2}\right) = \left(\frac{dy}{dt}\right)^2$$

Differential Equations from Physical Problems

Here are two more examples of applying DEs to physical problems.

Example 16.9

A stone is dropped from a cliff top and falls freely under gravity. Find the distance it has fallen after 3 seconds.

We have seen (on p. 159) that acceleration is denoted by $\frac{d^2s}{dt^2}$, where s is distance and t is time.

An object falling freely under gravity has a constant acceleration of $g = 9.81$ m/s^2.

In the form of a DE then
$$\frac{d^2s}{dt^2} = g \qquad [1]$$

Integrating w.r.t. t
$$\frac{ds}{dt} = gt + A \qquad [2]$$

and again
$$s = \tfrac{1}{2}gt^2 + At + B \qquad [3]$$

Note that each time we integrate a constant of integration is introduced — first A and then B.

Now Equation [2] is a first order DE expressing the connection between velocity $\frac{ds}{dt}$ and time t.

Equation [3] is the general solution and gives distance s in terms of time t. Now if we take the top of the cliff as our datum (starting point) then:

(a) the stone starts from rest at the cliff top. In other words its velocity is zero at zero time, which may be stated as $\frac{ds}{dt} = 0$ when $t = 0$

(b) the distance fallen is zero at zero time, or $s = 0$ when $t = 0$

Now these are two sets of boundary conditions which may be used to find the particular solution of the DE.

Substituting $\frac{ds}{dt} = 0$ when $t = 0$ in Equation [2] $0 = g(0) + A$
$$\therefore A = 0$$

Substituting $s = 0$ when $t = 0$ in Equation [3] $0 = \tfrac{1}{2}g(0)^2 + A(0) + B$
$$\therefore B = 0$$

So the particular solution is $s = \tfrac{1}{2}gt^2$

When $t = 3$ seconds, $s = \tfrac{1}{2}(9.81)(3)^2$
$$= 44.1 \text{ metres}$$

Example 16.10

The instantaneous rate of growth of a unit-sized population is proportional to the population size. Form a differential equation to express this fact, and then find a solution. If the population size is 20 after 4 hours from the commencement of growth find the time when the population size is 10.

Let the population size be N and the time t.

Now *rate of growth* is proportional to *population size*

or
$$\frac{dN}{dt} = kN \qquad \text{where } k \text{ is a constant.}$$

Rearranging,
$$\frac{dt}{dN} = \frac{1}{k}\left(\frac{1}{N}\right)$$

Integrating w.r.t. N,
$$t = \frac{1}{k}\int \frac{1}{N}\, dN$$

∴
$$t = \frac{1}{k}(\log_e N) + c \qquad\qquad [1]$$

This is the required solution. There are two unknown constants: k was in the original DE and c is a constant of integration. We have two sets of boundary conditions:

(a) unit population size at zero time giving $N = 1$ when $t = 0$
(b) population size of 20 after 4 hours giving $N = 20$ when $t = 4$

Substituting $N = 1$ when $t = 0$ in Equation [1],

$$0 = \frac{1}{k}(\log_e 1) + c$$

or
$$0 = \frac{1}{k}(0) + c$$

∴
$$c = 0$$

Substituting $N = 20$ when $t = 4$ in Equation [1]

$$4 = \frac{1}{k}(\log_e 20)$$

∴
$$k = \frac{1}{4}(\log_e 20)$$

from which
$$k = 0.749$$

Thus Equation [1] becomes
$$t = \frac{1}{0.749}(\log_e N)$$

When $N = 10$
$$t = \frac{1}{0.749}(\log_e 10) = 3.07 \text{ hours.}$$

Note: The above is an alternative way of solving any equation of the type $\dfrac{dy}{dx} = ky$, which was solved by another method on p. 235

Example 16.11

In a chemical reaction of the second order, the rate at which the concentration decreases is directly proportional to the square of the concentration. If x is the concentration at time t and the constant of proportionality is k_r (known as the rate constant in chemical reactions), form a DE to represent the statement. Hence show that, if the initial concentration is a, one form of the solution is

$$t = k_r \left(\frac{1}{x} - \frac{1}{a} \right).$$

If the half-life occurs after time of 152, for an initial concentration of 0.52, find the value of the rate constant k_r.

The DE, in which the negative sign represents decrease (as opposed to increase), will be

$$\frac{dx}{dt} = -k_r x^2$$

Rearranging,

$$\frac{dt}{dx} = -\frac{1}{k_r x^2}$$

Integrating w.r.t. x,

$$t = -\frac{1}{k_r} \int x^{-2} \, dx$$

$$= -\frac{1}{k_r} \left(\frac{x^{-1}}{-1} \right) + A$$

or

$$t = \frac{1}{k_r x} + A$$

Now the initial concentration is a, which means that $x = a$ when $t = 0$, and substituting these values we have

$$0 = \frac{1}{k_r a} + A$$

from which

$$A = -\frac{1}{k_r a}$$

Thus the solution becomes

$$t = \frac{1}{k_r x} - \frac{1}{k_r a}$$

or

$$t = \frac{1}{k_r} \left(\frac{1}{x} - \frac{1}{a} \right)$$

The half-life of the reaction is the time required for the initial concentration a to decrease to half its value — that is $\frac{a}{2}$.

Thus   Half-life time $= \dfrac{1}{k_r}\left(\dfrac{1}{a/2} - \dfrac{1}{a}\right)$

$$= \dfrac{1}{k_r}\left(\dfrac{2}{a} - \dfrac{1}{a}\right)$$

$$= \dfrac{1}{k_r a}$$

So if the half-life time $= 152$ and the initial concentration $a = 0.52$ then

$$152 = \dfrac{1}{k_r(0.52)}$$

giving $$k_r = \dfrac{1}{(152)(0.52)}$$

$$= 0.0127$$

EXERCISE

16·4

1 The equation $y = x^2 + bx$ represents a family of parabolas. To represent this relationship, independent of constant b, find

(a) a second order DE   (b) a first order DE

2 If $y = x + \dfrac{c}{x}$ form a first order DE independent of constant c.

3 For $y = Ax^2 + Bx$ form a second order DE independent of constants A and B.

4 For $y = Ax^3 + Bx$ form a second order DE independent of constants A and B.

5 Displacement y and time t in simple harmonic motion (SHM) are related by the equation $y = a \sin \omega t$. Represent this by a second order DE which is independent of constant a.

6 If $\theta = a \sin t + b \cos t$ find a second order DE, independent of constants a and b, to represent his relationship.

7 A bullet is fired vertically upwards with an initial velocity u. Form a DE for its acceleration and by integrating twice find an equation for the distance s travelled by the bullet in terms of the time t from firing.

8 In a chemical reaction of the first order the rate at which a concentration decreases is directly proportional to the concentration. If x is the concentration at time t, and the constant of proportionality is k_r (known as the rate constant in chemical reactions), form a DE to represent this statement. Hence show that if the initial concentration is a, then one form of the solution of the DE is

$$t = \frac{1}{k_r} \log_e \left(\frac{a}{x} \right)$$

If the half life is 5.89×10^3, find the value of the rate constant k_r.

9 In a chemical reaction of the third order the rate at which a concentration decreases is directly proportional to the cube of the concentration. If x is the concentration at time t and the constant of proportionality k_r (known as the rate constant in chemical reactions), form a DE to represent this statement. Hence show that if the initial concentration is a then one form of the solution of the DE is

$$t = \frac{1}{2k_r} \left(\frac{1}{x^2} - \frac{1}{a^2} \right)$$

Show also that the half life is given by the expression $\dfrac{3}{2k_r a^2}$

Solution of Differential Equations by Separating the Variables

Consider the DE

$$y \frac{dy}{dx} = 3x^2$$

Here we are allowed to separate the dx and dy terms and rearrange to give

$$y \, dy = 3x^2 \, dx$$

We have gathered all the y terms on one side of the equation, and all the x terms on the other. This process is called **separating the variables**.

We now integrate both sides of the equation and then continue until we arrive at the general solution.

Thus

$$\int y \, dy = \int 3x^2 \, dx$$

giving

$$\frac{y^2}{2} + c = 3\frac{x^3}{3} + k$$

or

$$y^2 = 2x^3 + 2(k - c)$$

For simplicity we replace $2(k - c)$ by another constant A. In future remember that only one constant is necessary in this type of solution. In solving DEs symbols A, B, C, etc. are more popular than c or k when representing constants.

Hence

$$y^2 = 2x^3 + A$$

or

$$y = \sqrt{2x^3 + A}$$

which is the general solution of the original DE.

Example 16.12

Find the general solution of $\dfrac{dy}{dx} = 5x + 6x^2$.

We may integrate this directly (as in Example 16.3) but it can also be solved by separating the variables.

So rearranging, $1 \, dy = (5x + 6x^2) \, dx$

Integrating, $\displaystyle\int 1 \, dy = \int (5x + 6x^2) \, dx$

∴

$$y = 5\frac{x^2}{2} + 6\frac{x^3}{3} + A$$

or

$$y = 2.5x^2 + 2x^3 + A$$

which is the general solution.

Example 16.13

Find the general solution of $\dfrac{dx}{dt} = x$ and the particular solution if $x = 1$ when $t = 2$.

An alternative method of solving a DE of this type was used in Example 16.10 where a solution was found expressing the equivalent of t in terms of x. Here we shall find the true general solution giving x in terms of t.

Separating the variables, $\qquad \dfrac{1}{x}\, dx = 1\, dt$

Integrating, $\qquad \displaystyle\int \frac{1}{x}\, dx = \int 1\, dt$

$\therefore \qquad \log_e x = t + A$

Rearranging, $\qquad x = e^{(t+A)} \quad$ since if $\ \log_b N = x$

$\qquad\qquad\qquad\qquad\qquad\qquad$ then $\qquad N = b^x$

This is the general solution, but it may be rearranged to obtain a 'neater' solution as follows

$$x = e^t \times e^A \quad \text{since } x^m \times x^n = x^{(m+n)}$$

or $\qquad x = e^A \times e^t$

and if we now replace constant e^A with another constant B then we have the general solution as

$$x = Be^t$$

Now $x = 1$ when $t = 2$.

$\therefore \qquad 1 = Be^2$

from which $\qquad B = \dfrac{1}{e^2} = 0.135$

Thus the particular solution is

$$x = 0.135e^t$$

Example 16.14

Find the general solution of $\dfrac{dy}{dx} = \dfrac{y+1}{x}$.

Separating the variables, $\qquad \dfrac{1}{y+1}\, dy = \dfrac{1}{x}\, dx$

Integrating, $\qquad \displaystyle\int \frac{1}{y+1}\, dy = \int \frac{1}{x}\, dx$

$\left(\text{see Example 14.20 for } \displaystyle\int \frac{dy}{y+1} \right)$

$\therefore \qquad \log_e(y+1) = \log_e x + A$

We now make the 'clever' move of replacing constant A by constant $\log_e B$ giving

$$\log_e(y + 1) = \log_e x + \log_e B$$

or $\qquad\qquad\qquad \log_e(y + 1) = \log_e(xB)$

since $\quad \log m + \log n = \log mn$

Thus $\qquad\qquad\qquad\qquad y + 1 = xB$

giving the general solution $\qquad y = Bx - 1$

Example 16.15

Find the general solution of $e^{2y} \dfrac{dy}{dx} = \cos x$ and the particular solution if $x = 0.5$ when $y = 0.1$.

Separating the variables $\qquad e^{2y}\, dy = (\cos x)\, dx$

Integrating, $\qquad\qquad\qquad \displaystyle\int e^{2y}\, dy = \int (\cos x)\, dx$

$\therefore \qquad\qquad\qquad\qquad \tfrac{1}{2}e^{2y} = \sin x + A$

or $\qquad\qquad\qquad\qquad e^{2y} = 2 \sin x + 2A$

but we need y in terms of x, so we shall rewrite this equation in log form giving

$$2y = \log_e(2 \sin x + 2A)$$

and if we replace constant $2A$ with constant B then we have the general solution

$$y = \tfrac{1}{2}\log_e(2 \sin x + B)$$

Now $x = 0.5$ when $y = 0.1$.

$\therefore \qquad\qquad\qquad 0.1 = \tfrac{1}{2}\log_e(2 \sin 0.5 + B)$

or $\qquad\qquad\qquad 0.2 = \log_e(0.959 + B)$

or $\qquad\qquad\qquad e^{0.2} = 0.959 + B$

from which $\qquad\qquad B = 1.221 - 0.959 = 0.262$

Thus the particular solution is $\qquad y = \tfrac{1}{2}\log_e(2 \sin x + 0.262)$

EXERCISE

16·5

Find the general solutions of the following DEs:

1 $\dfrac{dy}{dx} = 3x^2 + 6x$ **2** $x\dfrac{dy}{dx} = x^3 + 6$

3 $\dfrac{dy}{dx} = \dfrac{1}{y}$ **4** $\dfrac{dy}{dx} = \dfrac{2+x}{2y}$

5 $\dfrac{dy}{dx} = \sin 3x + \cos 2x$ **6** $\dfrac{dy}{dx} = e^{2x} + 6$

7 $\dfrac{dy}{dx} = \dfrac{x}{y}$ **8** $\dfrac{dy}{dx} = xy$

9 $\dfrac{dy}{dx} = \dfrac{y}{x}$ **10** $y\dfrac{dy}{dx} = \dfrac{x+1}{y}$

11 $\cos\theta\,\dfrac{d\theta}{dt} = t$ **12** $\dfrac{dy}{dx} = \dfrac{x(1+x^2)}{y^2}$

Find the particular solutions of the following DEs:

13 $\dfrac{dy}{dx} = \dfrac{\sin x}{y}$ if $x = 0.5$ when $y = 1$

14 $e^x\dfrac{dy}{dx} = 3$ if $x = 0$ when $y = 2$

15 $y\dfrac{dy}{dx} = \sec^2 x$ if $x = 2$ when $y = 1$

16 $\dfrac{dy}{dx} = \dfrac{x+1}{x}$ if $x = 2$ when $y = 0$

17 $\cos x\,\dfrac{dy}{dx} = -\sin x$ if $x = 0.5$ when $y = 0.1$

18 $\dfrac{dy}{dx} = e^{(x-y)}$ if $x = 0$ when $y = 1$

19 $v\dfrac{dv}{dt} = 6t + 7$ if $v = 1$ when $t = 0$

20 $(u^2 + 1)\dfrac{dv}{du} = v^2$ if $u = 0$ when $v = \frac{1}{3}$

Centroids of Areas and Volumes

OBJECTIVES

1 Sketch a given area including a typical incremental area whose centroid is known.

2 Determine the moment of the increment in **1** about a specified axis in the plane of the area.

3 Sum the moments of the increments between given limits by definite integration.

4 Determine the given area.

5 Define centroid as the ratio of the total moment determined in **3** to the total area determined in **4**.

6 Calculate the distance of the centroid from the given axis.

7 Calculate the centroids of a rectangle, triangle, circle and sector of a circle.

8 Verify the position of the centroid of a semicircle using a theorem of Pappus.

9 Determine a volume by summing suitable incremental volumes using definite integration.

10 Determine the position of the centroid of volumes using definite integration.

Centroids of Areas

We know that the moment of a force F about the line AB (Fig. 17.1) is given by $F \times d$.

Fig. **17.1**

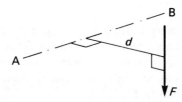

Similarly we say that the first moment of the area A about the line CD (Fig. 17.2) is given by $A \times d$, the point G being the centroid of the area.

Fig. **17.2**

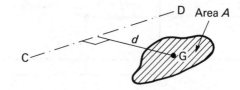

The centroid of an area is the point at which the total area may be considered to be situated for calculation purposes.

The centroid of an area is at the point which corresponds to the centre of gravity of a lamina of the same shape as the area. A thin flat sheet of steel of uniform thickness is an example of a lamina.

We know the position of the centres of gravity of the commonest shapes of laminae: for example the centre of gravity of a thin flat circular disc is at the centre of the disc. Hence we may say that the centroid of a circular area is at the centre of the area.

It is often possible to deduce the position of the centroid by the symmetry of the area. For example, in the case of a rectangle the centroid will be at the intersection of the centre lines, as shown in Fig. 17.3, as the area is symmetrical about either of the centre lines.

Fig. **17.3**

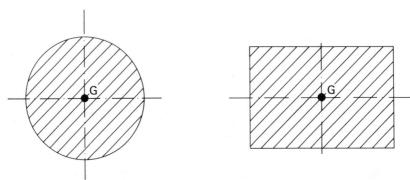

Positions of the centroid G for a circular and a rectangular area.

It is often necessary to find the position of the centroid of a *composite area*. This is an area which is made up from common shapes. To determine the position of the centroid we use two reference axes Ox and Oy. The location of the centroid is then given by the co-ordinates $\bar{x}$ and $\bar{y}$ as shown in Fig. 17.4.

Fig. **17.4**

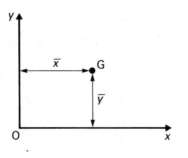

To find $\bar{x}$ and $\bar{y}$ we use the following formulae, which you have met previously:

$$\bar{x} = \frac{\Sigma Ax}{\Sigma A} \quad \text{and} \quad \bar{y} = \frac{\Sigma Ay}{\Sigma A}$$

where A is the sum of the component areas

whilst ΣAx and ΣAy are the sum of the first moments of the component areas about the axes Oy and Ox respectively.

Example 17.1

Find the position of the centroid shown in Fig. 17.5.

Fig. **17.5**

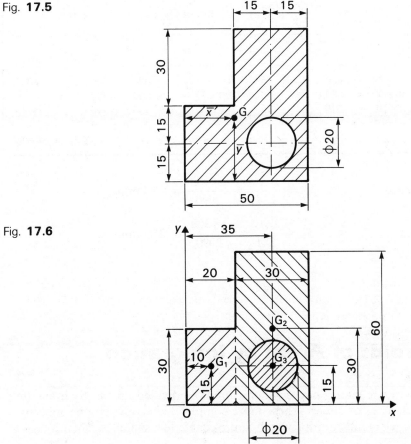

Fig. **17.6**

For convenience the reference axes Ox and Oy have been chosen as shown in Fig. 17.6. It will also be seen that the given area has been divided into three component areas whose centroids are G_1, G_2 and G_3.

The dimensions of each component area and the location of each centroid from the axes Ox and Oy are clearly shown. A diagram of this type is essential for this and similar problems — any attempt to obtain these dimensions mentally usually results in one or more errors. Remember the circle is a 'missing' area and must be taken as negative in the calculations.

It helps to simplify the arithmetic and hence reduce errors by showing the solution in tabular form.

Area	Distance to centroid		Moment of area	
	from Oy	from Ox	about Oy	about Ox
A	x	y	Ax	Ay
$20 \times 30 = 600$	10	15	$600 \times 10 = 6\,000$	$600 \times 15 = 9\,000$
$30 \times 60 = 1800$	35	30	$1800 \times 35 = 63\,000$	$1800 \times 30 = 54\,000$
$-\pi \times 10^2 = -310$	35	15	$-310 \times 35 = -10\,900$	$-310 \times 15 = -4700$
$\Sigma A = 2090$			$\Sigma Ax = 58\,100$	$\Sigma Ay = 58\,300$

Hence

$$\bar{x} = \frac{\Sigma Ax}{\Sigma A} = \frac{58\,100}{2090} = 27.8 \text{ mm}$$

and

$$\bar{y} = \frac{\Sigma Ay}{\Sigma A} = \frac{58\,300}{2090} = 27.9 \text{ mm}$$

Centroids of Areas by Integration

There are many shapes which cannot be divided up exactly into either rectangles or circles. These more complicated shapes can be split up into elementary strips which approximate to rectangles. Then by finding the dimensions of each strip etc. the formulae $\bar{x} = \dfrac{\Sigma Ax}{\Sigma A}$ and $\bar{y} = \dfrac{\Sigma Ay}{\Sigma A}$ may be used to find the approximate values of $\bar{x}$ and $\bar{y}$.

However, if we know, or can find, the equation of y in terms of x, by setting up the area on suitable axes, the numerical summing up may be achieved by integration and exact results obtained.

The following examples will explain how integration is used to find a centroid.

Example 17.2

Find, by integration, the distance of the centroid from the left-hand edge for the rectangle shown in Fig. 17.7.

Fig. **17.7**   Fig. **17.8**

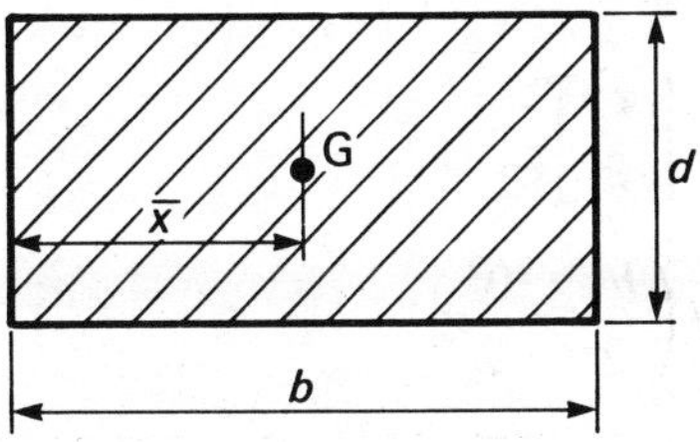

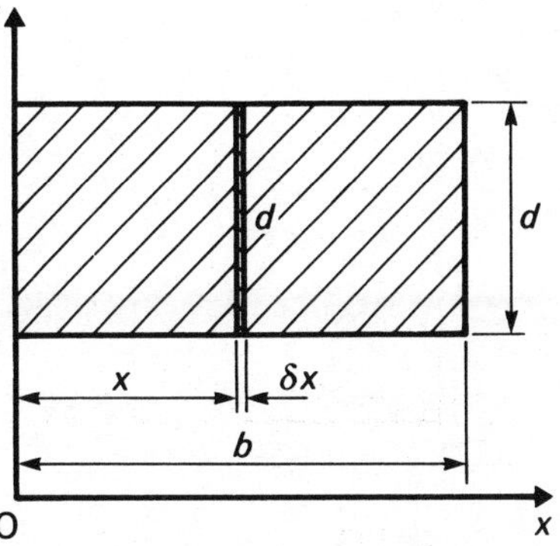

The rectangle may be set up on suitable axes. Fig. 17.8 shows a sketch of the arrangement and includes the usual elementary strip of very small width δx and length d. We now need to find distance $\bar{x}$.

Now the area of the rectangle $= \Sigma A$

$$= \text{Sum of the elementary strip areas}$$

$$= \sum_{x=0}^{x=b} d\,\delta x$$

$$= \int_0^b d\,\mathrm{d}x$$

$$= d\int_0^b 1\,\mathrm{d}x$$

$$= d\,[x]_0^b = d(b-0) = bd$$

Also the first moment of the rectangular area about the y-axis

$$= \Sigma Ax$$

$$= \text{Sum of the first moment of area of each of the elementary strips about } Oy$$

$$= \text{Sum of: (area of strip} \times \text{distance of its centroid from the } y\text{-axis})$$

$$= \sum_{x=0}^{x=b} (d\delta x)x$$

$$= \int_0^b d \, dx \, x$$

$$= d \int_0^b x \, dx$$

$$= d \left[\frac{x^2}{2} \right]_0^b$$

$$= d \left(\frac{b^2}{2} - \frac{0^2}{2} \right)$$

$$= \frac{b^2 d}{2}$$

Hence

$$\bar{x} = \frac{\Sigma Ax}{\Sigma A} = \frac{b^2 d/2}{bd} = \frac{b}{2}$$

This is the result we should expect from the symmetry of the figure. The procedure followed in this example is always used when finding positions of centroids by integration.

Example 17.3

Find the position of the centroid of a triangle of height H and base length B.

The triangle must be set up on suitable axes. Only experience enables a good choice. In this case it is convenient to turn the triangle through $90°$ so that the apex lies on the vertical y-axis and the base is parallel to it. In this instance the position of the x-axis is unimportant.

Fig. **17.9**

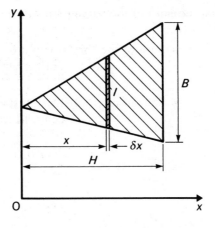

Fig. 17.9 shows a sketch of the arrangement and includes the usual elementary strip area of very small width δx and length l. We shall need a connection between l and x and this may be found using similar triangles giving $\dfrac{l}{x} = \dfrac{B}{H}$ from which $l = \dfrac{B}{H}x$.

Now the area of the triangle $= \Sigma A$

$\qquad\qquad\qquad\qquad\quad =$ Sum of the elementary strip areas

$$= \sum_{x=0}^{x=H} l\,\delta x$$

$$= \int_0^H l\,dx$$

$$= \int_0^H \frac{B}{H}x\,dx$$

$$= \frac{B}{H}\int_0^H x\,dx$$

$$= \frac{B}{H}\left[\frac{x^2}{2}\right]_0^H = \frac{B}{H}\left(\frac{H^2}{2} - \frac{0^2}{2}\right) = \tfrac{1}{2}BH$$

Also the first moment of the triangular area about the y-axis is

$\Sigma Ax =$ Sum of the first moment of area of each
of the elementary strips about Oy

$=$ Sum of (area of strip $\times$ distance of its
centroid from the y-axis)

$$= \sum_{x=0}^{x=H} l \, \delta x \, x$$

$$= \int_0^H l \, dx \, x$$

$$= \int_0^H \frac{B}{H} x \, dx \, x$$

$$= \frac{B}{H} \int_0^H x^2 \, dx$$

$$= \frac{B}{H} \left[\frac{x^3}{3} \right]_0^H = \frac{B}{H} \left(\frac{H^3}{3} - \frac{0^3}{3} \right) = \tfrac{1}{3} BH^2$$

Hence $$\bar{x} = \frac{\Sigma Ax}{\Sigma A} = \frac{\tfrac{1}{3} BH^2}{\tfrac{1}{2} BH} = \tfrac{2}{3} H$$

It should be noted that this is independent of the base length B.

You may remember that the position of the centroid of a triangular area is as given in Fig. 17.10. The above calculations verify this.

Fig. **17.10**

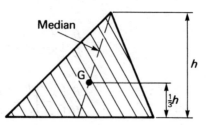

Median

h

$\tfrac{1}{3}h$

G

Example 17.4

Find the position of the centroid of the sector of a circle of radius R and angle 2α.

The sector area is shown in Fig. 17.11 located on suitable axes Ox and Oy. By symmetry the centroid G of the sector area lies on the x-axis, and we need to find the distance x. It is not convenient to divide the area into vertical (or even horizontal) elementary strip areas. Instead we consider elementary sector areas of small angle $\delta\theta$ (Fig. 17.12). Each of these elementary areas may be considered to approximate to a triangle having base length $R \times \delta\theta$ (since length of arc $=$ angle $\times$ radius, where the angle is in radians), and height R.

Fig. **17.11**   Fig. **17.12**

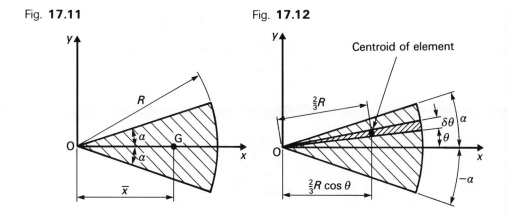

We must first find the area of the sector by summing the areas of the elementary sector areas.

Hence   Area of sector $= \Sigma A$

$\qquad\qquad\qquad\quad =$ Sum of elementary sector areas

$$= \sum_{\theta=-\alpha}^{\theta=\alpha} \tfrac{1}{2}(R\,\delta\theta)R \qquad \text{since the area of a triangle is } \tfrac{1}{2}(\text{base})(\text{height})$$

$$= \int_{-\alpha}^{\alpha} \tfrac{1}{2}(R\,\mathrm{d}\theta)R$$

$$= \tfrac{1}{2}R^2 \int_{-\alpha}^{\alpha} \mathrm{d}\theta = \tfrac{1}{2}R^2 \left[\theta\right]_{-\alpha}^{\alpha}$$

$$= \tfrac{1}{2}R^2 \{\alpha - (-\alpha)\}$$

$$= R^2\alpha$$

Now the first moment of the given area about the y-axis is

$$\Sigma Ax = \text{Sum of the first moment of area of each of the elementary sector areas about } Oy$$

$$= \text{Sum of (elementary area} \times \text{distance of its centroid from the } y\text{-axis)}$$

$$= \sum_{\theta = -\alpha}^{\theta = \alpha} (\tfrac{1}{2}R^2\delta\theta)(\tfrac{2}{3}R\cos\theta)$$

$$= \int_{-\alpha}^{\alpha} (\tfrac{1}{2}R^2\,\mathrm{d}\theta)(\tfrac{2}{3}R\cos\theta)$$

$$= \tfrac{1}{3}R^3 \int_{-\alpha}^{\alpha} \cos\theta\,\mathrm{d}\theta = \tfrac{1}{3}R^3\left[\sin\theta\right]_{-\alpha}^{\alpha}$$

$$= \tfrac{1}{3}R^3\{\sin\alpha - \sin(-\alpha)\}$$

$$= \tfrac{1}{3}R^3\{\sin\alpha + \sin\alpha\}$$

$$= \tfrac{2}{3}R^3\sin\alpha$$

Now

$$\bar{x} = \frac{\Sigma Ax}{\Sigma A}$$

$$= \frac{\tfrac{2}{3}R^3\sin\alpha}{R^2\alpha}$$

$$= \tfrac{2}{3}R\,\frac{\sin\alpha}{\alpha}$$

It should be remembered that the angle α must be in radians.

Example 17.5

Using the result obtained for the position of the centroid of the sector of a circle find:

(a) the position of the centroid of a quadrant of a circle of radius R
(b) the position of the centroid of a semicircular area of radius R

(a) If we set the quadrant area on suitable axes as shown in Fig. 17.13, then we may use the formula for the sector of a circle, giving

$$OG = \tfrac{2}{3}R\,\frac{\sin\alpha}{\alpha} \quad \text{where} \quad \alpha = \frac{\pi}{4}\text{ rad}$$

Fig. **17.13**

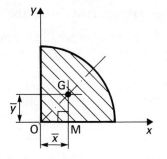

From right-angled triangle OGM we have

$$\bar{x} = \text{OM}$$

$$= \text{OG}\left(\cos\frac{\pi}{4}\right)$$

$$= \tfrac{2}{3}R\,\frac{\sin\dfrac{\pi}{4}}{\dfrac{\pi}{4}}\left(\cos\frac{\pi}{4}\right)$$

$$= \frac{2R(0.7071)(0.7071)4}{3\pi}$$

$$= \frac{4R}{3\pi}$$

$$= 0.424R$$

Also by symmetry $\bar{y}$ will have the same value as $\bar{x}$.

(b) The semicircular area shown in Fig. 17.14 can be considered to be two quadrants as indicated. It follows, therefore, that $\bar{x}$ for a semicircular area is the same as that for a quadrant, i.e. 0.424R.

Fig. **17.14**

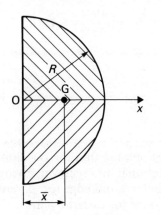

By symmetry the centroid also lies on the horizontal centre line Ox.

The above result may be confirmed by using a theorem by Pappus which you may remember, stating:

> If a plane area rotates about a line in its plane (which does not cut the area) then the volume generated is given by the equation
>
> Volume = Area × Length of path of its centroid

When a semicircular area is rotated about its boundary diameter a sphere is generated. Since the radius of the semicircle is R, the volume of the sphere generated is $\frac{4}{3}\pi R^3$.

Let the required distance be $\bar{x}$ in Fig. 17.15.

Fig. **17.15**

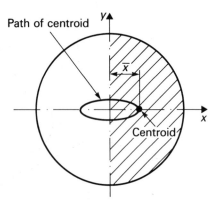

By Pappus' Theorem,

$$\text{Volume} = \text{Area} \times \text{Path of centroid of the area,}$$

$$\therefore \quad \tfrac{4}{3}\pi R^3 = \tfrac{1}{2}\pi R^2 \times 2\pi \bar{x}$$

$$\therefore \quad \bar{x} = \frac{4R}{3\pi} = 0.424R$$

EXERCISE

17·1   In Questions 1–4 find the positions of the centroids of the cross-sectional areas shown, giving the distances in each case from the left-hand edge (or extreme point) and above the bottom edge (or extreme point).

1

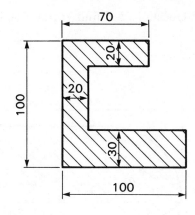

2

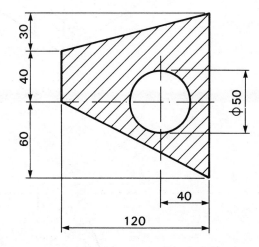

3

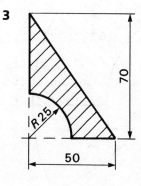

4

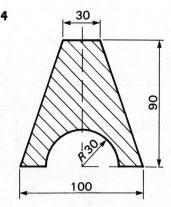

In Questions 5–8 find $\bar{x}$ and $\bar{y}$ as indicated on the diagrams.

5

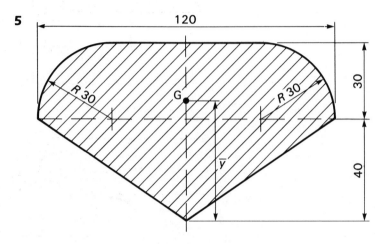

6

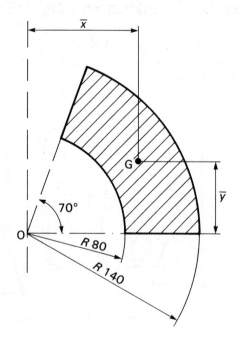

7

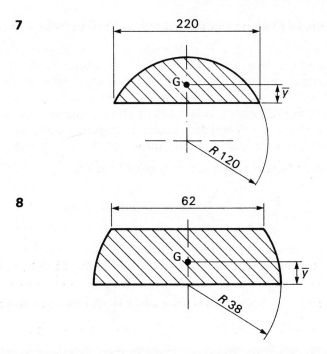

8

Volumes of Revolution by Integration

If the area under the curve APB (Fig. 17.16) is rotated one complete revolution about the x-axis, then the volume swept out is called a volume of revolution.

Fig. **17.16**

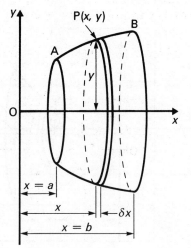

The point P, whose coordinates are (x, y), is a point on the curve AB.

Let us consider, below P, a thin slice whose width is δx. Since the width of the slice is very small we may consider the slice to be a cylinder of radius y. The volume of this slice is approximately $\pi y^2 \delta x$. Such a slice is called an elementary slice and we shall consider that the volume of revolution is made up from many such elementary slices. Hence the complete volume of revolution is the sum of all the elementary slices between the values of $x = a$ and $x = b$.

In mathematical notation this may be stated as

$$\sum_{x=a}^{x=b} \pi y^2 \delta x \quad \text{approximately.}$$

As for areas, the process of integration may be considered to sum an infinite number of elementary slices and hence it gives an exact result.

$$\therefore \qquad \text{Volume of revolution} = \int_a^b \pi y^2 \, dx \quad \text{exactly}$$

Example 17.6

The area between the curve $y = x^2$, the x-axis and the ordinates $x = 1$ and $x = 3$ is rotated about the x-axis. Find the volume of revolution.

As when finding areas it is recommended that a sketch is made of the required volume. Fig. 17.17 shows a sketch of the required volume.

Fig. **17.17**

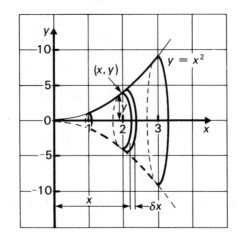

Required volume of revolution $= \int_1^3 \pi y^2 \, dx$

$$= \int_1^3 \pi (x^2)^2 \, dx$$

$$= \pi \int_1^3 x^4 \, dx$$

$$= \pi \left[\frac{x^5}{5} \right]_1^3$$

$$= \pi \left(\frac{3^5}{5} - \frac{1^5}{5} \right)$$

$$= 152 \text{ cubic units.}$$

Example 17.7

Find, by the calculus, the volume of a cone of base radius R and height H.

The first step is to set up the cone on suitable axes. For convenience the cone has been put with its polar axis lying along the x-axis as shown in Fig. 17.18.

Fig. **17.18**

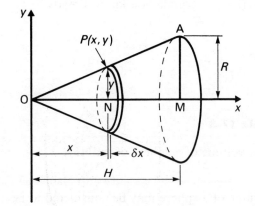

Then the volume of the cone is the volume of revolution generated when the area OAM is rotated about the x-axis.

We need an equation connecting x and y. This may be found by considering the similar triangles OPN and OAM.

Then
$$\frac{PN}{ON} = \frac{AM}{OM}$$

or
$$\frac{y}{x} = \frac{R}{H}$$

from which
$$y = \frac{R}{H}x$$

Now the required volume
$$= \int_0^H \pi y^2 \, dx$$

$$= \int_0^H \pi \left(\frac{R}{H}x\right)^2 dx$$

$$= \pi \frac{R^2}{H^2} \int_0^H x^2 \, dx$$

$$= \pi \frac{R^2}{H^2} \left[\frac{x^3}{3}\right]_0^H$$

$$= \pi \frac{R^2}{H^2} \left(\frac{H^3}{3} - \frac{0^3}{3}\right)$$

$$= \tfrac{1}{3}\pi R^2 H$$

which verifies a formula you know already.

Example 17.8

Find the volume of a sphere of radius R.

The volume of a sphere may be considered to be a volume of revolution generated by the revolution of a semicircular area about its boundary diameter. Suitable axes must now be chosen and these are shown in Fig. 17.19.

Fig. **17.19**

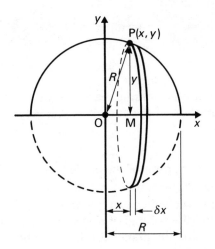

The relationship connecting y and x may be found by considering the right-angled triangle OPM and applying the theorem of Pythagoras.

Hence $\qquad x^2 + y^2 = R^2$

$\therefore \qquad\qquad y^2 = R^2 - x^2$

Now the volume of revolution $= \int \pi y^2 \, dx$

$$= \pi \int_{-R}^{R} (R^2 - x^2) \, dx$$

$$= \pi \left[R^2 x - \frac{x^3}{3} \right]_{-R}^{R}$$

$$= \pi \left\{ \left(R^2(R) - \frac{R^3}{3} \right) - \left(R^2(-R) - \frac{(-R)^3}{3} \right) \right\}$$

$$= \tfrac{4}{3}\pi R^3$$

verifying a formula you should recognise.

EXERCISE

17·2

In Questions 1–5 find the volume generated about the x-axis of the given curves between the limits stated.

1 $y = x^3$ from $x = 0$ to $x = 2$.

2 $xy = 16$ from $x = 1$ to $x = 2$.

3 $x = 4\sqrt{y}$ from $x = 1$ to $x = 2$.

4 $y^3 = x^2$ from $x = 1$ to $x = 8$.

5 $y = \dfrac{x}{2}$ from $x = 0$ to $x = 4$.

6 Find the volume generated by revolving about the x-axis the portion of the curve $y = x(x - 1)$ that lies below the x-axis.

7 The portion of the circle $x^2 + y^2 = 25$ in the first quadrant is rotated about the x-axis. Find:
(a) the volume of a hemisphere of radius 5 m
(b) the volume of a section of the above hemisphere cut off by the planes distant 2 m and 3 m from the plane base

8 A bucket has a radius of 100 mm at the base, and 200 mm at the top. It is 200 mm deep and the sides slope uniformly. Show that it may be considered as a volume of revolution about its axis of symmetry. Hence find the capacity of the bucket, when full, in litres.

9 Fig. 17.20 shows the cross-section through a brass nozzle. Find the volume of the nozzle.

Fig. **17.20**

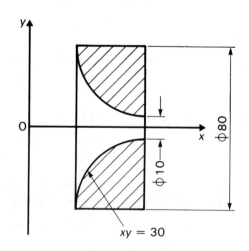

$xy = 30$

Centroids of Volumes by Integration

The centroid of a volume is at the point which corresponds to the centre of gravity of a uniform solid body having the same shape. We know the position of the centres of gravity of many solid bodies. For example the centre of gravity of a spherical mass is at its centre; hence we may say that the centroid of a spherical volume is at the centre of the volume.

To find centroids of areas we used the formulae

$$\bar{x} = \frac{\Sigma Ax}{\Sigma A} \quad \text{and} \quad \bar{y} = \frac{\Sigma Ay}{\Sigma A}$$

For centroids of volumes similar expressions, replacing area A by volume V, are used:

$$\bar{x} = \frac{\Sigma Vx}{\Sigma V} \quad \text{and} \quad \bar{y} = \frac{\Sigma Vy}{\Sigma V}$$

Example 17.9

Find the centroid of a right circular conical volume of height H and base radius R.

For convenience we shall set the cone on axes as shown in Fig. 17.21. By symmetry the centroid lies on the x-axis.

Fig. **17.21**

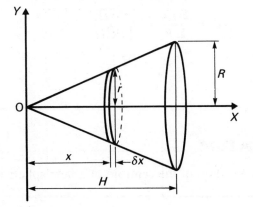

It remains to find the distance $\bar{x}$ of the centroid from the y-axis.
As when finding volumes we consider the cone to be made up from elementary slices, each of radius r and thickness δx, whose volumes are

$\pi r^2 \delta x$. The first moment Vx of this slice about the y-axis will be given by $(\pi r^2 \delta x)x$, and if we find the sum of these first moments we obtain the first moment of the whole conical volume. In mathematical notation this may be stated as

Sum of the first moments $\quad \Sigma \, Vx = \displaystyle\sum_{x=0}^{x=H} (\pi r^2 \delta x)x \quad$ approximately

$$= \int_0^H (\pi r^2 \, dx)x \quad \text{exactly}$$

We must now find r in terms of x. From similar right-angled triangles in Fig. 17.21 we have $\dfrac{r}{R} = \dfrac{x}{H}$ from which $r = \left(\dfrac{R}{H}\right)x$. Thus

$$\Sigma \, Vx = \int_0^H \left\{ \pi \left(\frac{R}{H}x\right)^2 dx \right\} x$$

$$= \frac{\pi R^2}{H^2} \int_0^H x^3 \, dx$$

$$= \frac{\pi R^2}{H^2} \left[\frac{x^4}{4} \right]_0^H$$

$$= \frac{\pi R^2}{H^2} \times \frac{H^4}{4}$$

$$= \tfrac{1}{4}\pi R^2 H^2$$

and since we know the volume of a cone is $\tfrac{1}{3}\pi R^2 H$,

$$\bar{x} = \frac{\Sigma \, Vx}{\Sigma \, V} = \frac{\tfrac{1}{4}\pi R^2 H^2}{\tfrac{1}{3}\pi R^2 H}$$

giving $\qquad\qquad\qquad \bar{x} = \tfrac{3}{4}H$

Example 17.10

Find the position of the centroid of a hemispherical volume of radius 80 mm.

Again for convenience we have set the hemisphere on axes as shown in Fig. 17.22 and consider the volume to comprise elementary slices as shown.

Fig. **17.22**

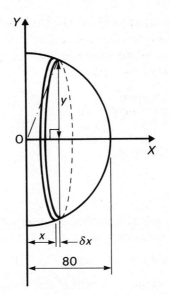

Thus $\qquad \Sigma Vx = \displaystyle\sum_{x=0}^{x=80} (\pi y^2 \, \delta x)x \quad$ approximately

$$= \int_0^{80} (\pi y^2 \, dx)x \quad \text{exactly}$$

Here we require y in terms of x and apply Pythagoras' theorem to the right-angled triangle in Fig. 17.22 which gives $\qquad 80^2 = x^2 + y^2$
from which $\qquad y^2 = 80^2 - x^2$

Thus $\qquad \Sigma Vx = \displaystyle\int_0^{80} \{\pi(80^2 - x^2) \, dx\}x$

$$= \pi \int_0^{80} (80^2 x - x^3) \, dx$$

$$= \pi \left[\frac{80^2 x^2}{2} - \frac{x^4}{4} \right]_0^{80}$$

$$= \pi \left(\frac{80^2 (80)^2}{2} - \frac{80^4}{4} \right)$$

$$= \pi \frac{80^4}{4}$$

But we know that the volume of a hemisphere of radius 80 is $\frac{2}{3}\pi(80)^3$.

Thus $\quad \bar{x} = \dfrac{\Sigma\, Vx}{\Sigma\, V}$

(Note that, by keeping the powers of 80, the final cancelling avoids much arithmetic!)

$$= \dfrac{\pi\,\dfrac{80^4}{4}}{\frac{2}{3}\pi(80)^3}$$

$$= \tfrac{3}{8}(80)$$

$$= 30 \text{ mm}$$

EXERCISE

Find, using integration, the position of the centroids along the axis of symmetry of the following volumes:

1 A cylinder of length L.

2 A pyramid of height H and square base.

3 A 1.5 m deep frustum of a cone, 1.2 m in diameter at the top and 2.4 m in diameter at the bottom.

4 A hemispherical dome with a radius of 4 m and having its top 1 m removed.

5 A 2 m deep cap of a 6 m radius sphere.

Second Moments of Area

18

OBJECTIVES

1 Sketch the given area including a typical incremental area parallel to a specified axis in the plane of the area.

2 Define the second moment of area.

3 Determine the second moment of the increment in **1** about the specified axis.

4 Determine the second moment of area by summing the second moment of the incremental area between given limits by definite integration.

5 Determine the second moment of a rectangle about one edge, a triangle about one edge, a circle and semicircle about a diameter, and a circle about its polar axis.

6 State the parallel axis theorem.

7 Determine the second moment of area about an axis through the centroid of the area parallel to the axis about which the second moment is known using **6**.

8 Determine the second moment of area about an axis parallel to the axis about which the second moment is known using **6** and **7**.

9 Calculate the second moments of the component areas of a composite area about a common axis using **8**.

10 Calculate the second moment of a composite area by summing the second moments in **9**.

11 State the perpendicular axis theorem.

12 Calculate the second moments of area of a given area about two axes at right angles in the plane of the given area.

13 Calculate the second moment of area about a perpendicular axis through the intersection of axes in **12** using **11**.

14 Define radius of gyration k from the expression $I = (\Sigma A)k^2$.

Second Moment of Area

The second moment of area is a property of an area used in many engineering calculations. One example is in finding stresses, due to bending, which involves the use of the second moment of the cross-sectional area of the beam.

Fig. 18.1 shows a thin strip, of area A and of very small width, which is distant x from the reference line CD.

Now the first moment of the area A about the reference line CD is given by Ax.

Similarly the second moment of area A about the reference line CD is given by Ax^2.

Fig. **18.1**

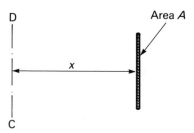

The symbol for the second moment of area is I and is always stated with reference to an axis or datum line. Thus for the strip area shown

$$I_{CD} = Ax^2$$

The expression Ax^2 is only true if the area is a very thin strip parallel to the reference line. It follows that if we require the second moment of any other shaped area it is necessary to divide the area into a number of elementary strips all parallel to the reference line.

Thus in Fig. 18.2 for the irregular area $I_{CD} = \Sigma Ax^2$.

Fig. **18.2**

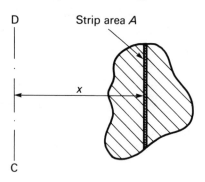

The summation of the Ax^2 for each strip may be carried out by graphical and numerical methods but, as in the case of finding first moments of area, this summation may often be achieved by integration as the following example will illustrate:

Example 18.1

Find the second moment of area of a rectangle of width b and depth d about its base edge.

The rectangular area must be set up on suitable axes. In this case it is convenient to turn the rectangle through 90° and arrange the base edge to lie on the y-axis as shown in Fig. 18.3. In this instance the position of the x-axis is unimportant.

Fig. **18.3**

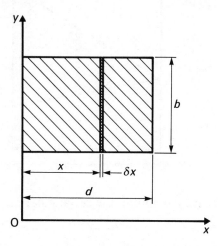

The diagram shows a typical elementary strip area parallel to the reference axis and whose area is $b\,\delta x$.

Now the second moment of the rectangular area about the y-axis

$$= \Sigma A x^2$$

$$= \sum_{x=0}^{x=d} (b\,\delta x)x^2$$

$$= \int_0^d b\,\mathrm{d}x\,x^2$$

$$= b \int_0^d x^2\,\mathrm{d}x$$

$$= b \left[\frac{x^3}{3} \right]_0^d$$

$$= \frac{bd^3}{3}$$

Example 18.2

Find the second moment of a triangular area, of height H and base B, about a line through its apex and parallel to the base.

The arrangement of the axes etc. is identical to that used in Example 17.3 and is as shown in Fig. 18.4.

Fig. **18.4**

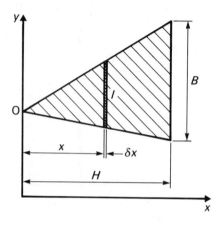

Now the second moment of the rectangular area about the y-axis

$$= \Sigma A x^2$$

$$= \sum_{x=0}^{x=H} (l\, \delta x) x^2$$

$$= \int_0^H l x^2 \, dx$$

and using similar triangles as before we have $\quad l = \dfrac{B}{H} x.$

Thus $\quad I_{OY} = \displaystyle\int_0^H \left(\frac{B}{H} x\right) x^2 \, dx$

$$= \frac{B}{H} \int_0^H x^3 \, dx$$

$$= \frac{B}{H} \left[\frac{x^4}{4}\right]_0^H = \frac{B}{H}\left(\frac{H^4}{4} - \frac{0^4}{4}\right) = \frac{B}{H}\left(\frac{H^4}{4}\right) = \frac{BH^3}{4}$$

Example 18.3

Find the second moment of a semicircular area, of radius R, about a boundary diameter.

For convenience the area will be set up on axes as shown in Fig. 18.5 and from the right-angled triangle we have

$$x = R \cos \theta \quad \text{and} \quad y = R \sin \theta$$

Fig. **18.5**

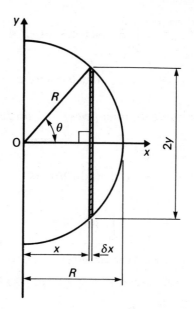

Now the second moment of area about the y-axis

$$I_{OY} = \Sigma A x^2$$

$$= \sum_{x=0}^{x=R} (2y\,\delta x)x^2$$

$$= 2\int_0^R y x^2\,\mathrm{d}x$$

$$= 2\int_0^R (R\sin\theta)(R\cos\theta)^2\,\mathrm{d}x$$

Now we have $\qquad x = R\cos\theta$

giving $\qquad \dfrac{\mathrm{d}x}{\mathrm{d}\theta} = -R\sin\theta$

or $\qquad \mathrm{d}x = -R\sin\theta\,\mathrm{d}\theta$

Also we change the limits:

When $x = 0$ then $\quad 0 = R\cos\theta$

giving $\qquad\qquad \theta = \dfrac{\pi}{2}$

When $x = R$ then $\quad R = R\cos\theta$

or $\quad 1 = \cos\theta$

giving $\qquad\qquad \theta = 0$

Thus $\qquad I_{OY} = 2 \int_{\pi/2}^{0} R^3 \sin\theta \cos^2\theta(-R\sin\theta\,d\theta)$

$$= -2R^4 \int_{\pi/2}^{0} (\sin\theta\cos\theta)^2\,d\theta$$

Now a double-angle formula gives $\qquad \sin 2\theta = 2\sin\theta\cos\theta$

from which $\qquad\qquad\qquad\qquad \sin\theta\cos\theta = \tfrac{1}{2}\sin 2\theta$

Hence $\qquad I_{OY} = -2R^4 \int_{\pi/2}^{0} (\tfrac{1}{2}\sin 2\theta)^2\,d\theta$

$$= -\frac{R^4}{2} \int_{\pi/2}^{0} \sin^2 2\theta\,d\theta$$

Another double-angle formula gives $\qquad \cos 4\theta = 1 - 2\sin^2 2\theta$

from which $\qquad\qquad\qquad\qquad \sin^2 2\theta = \tfrac{1}{2}(1 - \cos 4\theta)$

Thus $\qquad I_{OY} = -\frac{R^4}{2} \int_{\pi/2}^{0} \tfrac{1}{2}(1 - \cos 4\theta)\,d\theta$

$$= -\frac{R^4}{4} \left[\theta - \frac{\sin 4\theta}{4} \right]_{\pi/2}^{0}$$

$$= -\frac{R^4}{4} \left\{ \left(0 - \frac{\sin 0}{4}\right) - \left(\frac{\pi}{2} - \frac{\sin 2\pi}{4}\right) \right\}$$

$$= \frac{\pi R^4}{8}$$

For a circular area we can double the result obtained above and it is usually stated in terms of the diameter D. So if we put $R = \dfrac{D}{2}$ then

$$I_{\text{diameter}} = 2\left(\frac{\pi}{8}\right)\left(\frac{D}{2}\right)^4$$

$$= \frac{\pi D^4}{64}$$

Units of Second Moments of Area

The second moment of area is ΣAx^2. If all lengths are in metres, then the result will be $m^2 \times m^2 = m^4$.

For example in the formula $\dfrac{bd^3}{3}$ the units are $m \times m^3 = m^4$.

In typical engineering problems it is unlikely that dimensions of a cross-sectional area will be in metres. It is more probable that the units will be in millimetres, and if these are used the second moment of area will be in mm^4 which will result in large numbers. In practice it is usual to calculate second moments of area in cm^4. This makes the arithmetic as simple as possible.

This is one of the few times when units not recommended by the Système International (SI) are used in engineering calculations. In British industry producers of rolled steel sections used for beams and columns, etc. agreed with continental manufacturers to use cm^4 as units of second moments of area.

The second moment of the rectangular area shown in Fig. 18.6 about its base, or more correctly about an axis BB on which the base lies, is

$$I_{BB} = \frac{bd^3}{3} = \frac{3 \times 2^3}{3} = 8 \, cm^4$$

Fig. **18.6**

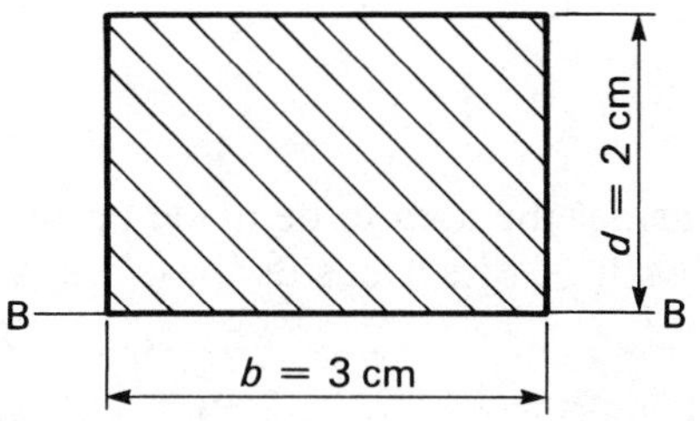

Conventional Notation I_{XX} and I_{YY}

XX and YY are known conventionally as the horizontal and vertical axes that pass through the centroid of a cross-sectional area.

Hence: I_{XX} is the second moment of a cross-sectional area about the horizontal axis that passes through the centroid,

and: I_{YY} is the second moment of a cross-sectional area about the vertical axis that passes through the centroid.

Example 18.4

Find I_{XX} and I_{YY} of the cross-sectional area shown in Fig. 18.7.

Fig. **18.7**

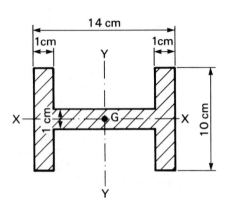

In this case by symmetry the centroid G will be at the centre of the section.

To find I_{XX}

We shall consider the area to be made up of six rectangles, as shown in Fig. 18.8, each of which has its 'base' on XX.

Fig. **18.8**

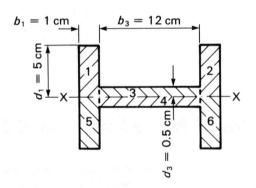

Now I_{XX} for the cross-sectional area will be the sum of the second moments of area of each rectangle about XX.

Rectangles 1, 2, 5 and 6 have the same dimensions and hence the same I_{XX}. Similarly rectangles 3 and 4 have the same I_{XX}.

Hence I_{XX} for the whole area $= 4(I_{XX}$ for area 1$) + 2(I_{XX}$ for area 3$)$

$$= 4\left(\frac{b_1 d_1^3}{3}\right) + 2\left(\frac{b_3 d_3^3}{3}\right)$$

$$= 4\left(\frac{1 \times 5^3}{3}\right) + 2\left(\frac{12 \times 0.5^3}{3}\right)$$

$$= 167.7 \text{ cm}^4$$

To find I_{YY}

Fig. **18.9**

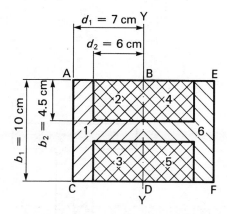

As before we shall consider the area to be made up of six rectangles, as shown in Fig. 18.9, each of which has its 'base' on YY. In this case the 'bases' will be vertical. The portion to the left of YY is area 1 (ABDC) less areas 2 and 3. This means that the I_{YY} of both areas 2 and 3 must be taken as negative. We treat the portion of the cross-section to the right of YY in a similar manner.

Rectangles 1 and 6 have the same dimensions and hence the same I_{YY}. Similarly rectangles 2, 3, 4 and 5 have the same I_{YY}.

Hence I_{YY} for the whole area $= 2(I_{YY}$ for area 1$) - 4(I_{YY}$ for area 2$)$

$$= 2\left(\frac{b_1 d_1^3}{3}\right) - 4\left(\frac{b_2 d_2^3}{3}\right)$$

$$= 2\left(\frac{10 \times 7^3}{3}\right) - 4\left(\frac{4.5 \times 6^3}{3}\right)$$

$$= 991 \text{ cm}^4$$

EXERCISE

Find I_{XX} and I_{YY} for the cross-sectional areas given in the following examples. All dimensions are in cm.

1

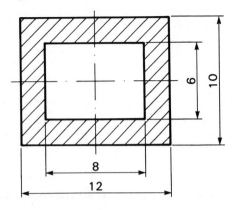

2

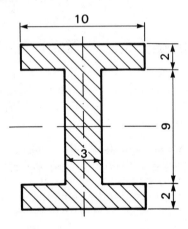

3

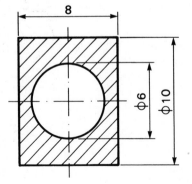

4

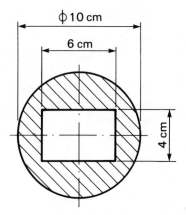

The Parallel Axis Theorem

A second moment of area must always be stated together with the axis about which it has been calculated. We state by saying, for example, that I_{AB} is the second moment of area about the axis AB.

Fig. 18.10 shows a plane area A whose centroid is G. Also shown is an axis that passes through G and a parallel axis, the distance between the axes being h.

Fig. **18.10**

The parallel axis theorem states:

$$I_{\text{parallel axis}} = I_{\text{axis through G}} + Ah^2$$

This is extremely useful as the following examples will show. It is worth while remembering that $I_{\text{axis through G}}$ is the numerically smallest second moment of area.

Example 18.5

Find the second moment of area of the rectangle shown in Fig. 18.11 about the axis XX which passes through the centroid.

Fig. **18.11**

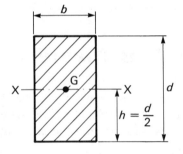

We know from Example 18.1 that the second moment of area of the rectangle about its base is $\dfrac{bd^3}{3}$.

Now using the parallel axis theorem we have

$$I_{\text{base}} = I_{XX} + Ah^2$$

$\therefore$ rearranging

$$I_{XX} = I_{\text{base}} - Ah^2$$

$$= \frac{bd^3}{3} - bd\left(\frac{d}{2}\right)^2$$

$$= bd^3\left(\frac{1}{3} - \frac{1}{4}\right)$$

$$= \frac{bd^3}{12}$$

Example 18.6

Given that the second moment of a triangular area, of height H and base B, about a line through its apex and parallel to its base is $\dfrac{BH^3}{4}$, find the second moment of the triangular area about its base.

Fig. 18.12 shows a sketch of the triangular area with axes suitably labelled. We are given I_{MM} and we require I_{NN}.

Fig. **18.12**

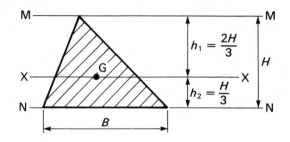

The parallel axis theorem requires that one of the axes used must pass through the centroid. It is necessary, therefore, to use the parallel axis theorem twice — once to find I_{XX} knowing I_{MM} and again to find I_{NN} using I_{XX}.

Now the parallel axis theorem gives $I_{MM} = I_{XX} + Ah_1{}^2$

and rearranging $I_{XX} = I_{MM} - Ah_1{}^2$

$\therefore$

$$I_{XX} = \frac{BH^3}{4} - \frac{BH}{2}\left(\frac{2H}{3}\right)^2$$

$$= \frac{BH^3}{36}$$

Also the parallel axis theorem gives $I_{NN} = I_{XX} + Ah_2{}^2$

$$= \frac{BH^3}{36} + \frac{BH}{2}\left(\frac{H}{3}\right)^2$$

$$= \frac{BH^3}{12}$$

Hence the second moment of a triangular area about its base is $\dfrac{BH^3}{12}$.

Summary of Second Moments of Common Areas

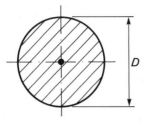

Circle

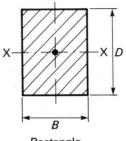

Rectangle

$$I_{\text{diameter}} = \frac{\pi D^4}{64}$$

$$I_{\text{XX}} = \frac{BD^3}{12}$$

$$I_{\text{base}} = \frac{BD^3}{3}$$

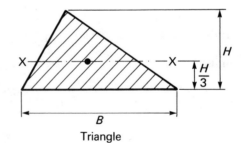

Triangle

$$I_{\text{XX}} = \frac{BH^3}{36}$$

$$I_{\text{base}} = \frac{BH}{12}$$

In practice most cross-sectional areas may be divided up into a combination of the above shapes — if not exactly, a close approximation can be achieved.

General Procedure for Finding I_{XX} and I_{YY} for Non-symmetrical Areas

Set up convenient vertical and horizontal axes, OV and OH. Divide the given area into suitable component areas and show them on a new diagram. This should also show the dimensions of each component area and the distances of the respective centroids from axes OV and OH.

To find I_{YY} for the given cross-sectional area:

(1) Find the position of the centroid using $\bar{x} = \dfrac{\Sigma Ax}{\Sigma A}$

(2) For each component area find:

    (a) I_{YY}^{G} about a vertical axis through its centroid G

    (b) Ax^2, x being the distance between G and OV

    (c) I_{OV} from $I_{OV} = I_{YY}^{G} + Ax^2$, using the parallel axis theorem.

(3) Find ΣI_{OV}, the second moment of the given cross-sectional area about OV.

(4) Find I_{YY}, the second moment of the given cross-sectional area about YY, using $I_{YY} = \Sigma I_{OV} - (\Sigma A)(\bar{x})^2$ from the parallel axis theorem.

To find I_{XX} use a similar sequence to that for finding I_{YY}.

It helps to simplify the arithmetic and hence reduce errors by using a tabular form for solutions as shown in the following examples:

Example 18.7

Find I_{XX} and I_{YY} for the cross-sectional area shown in Fig. 18.13. All dimensions are in cm.

Fig. **18.13**

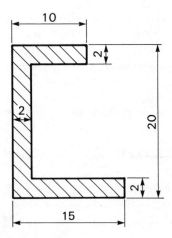

Fig. **18.14**

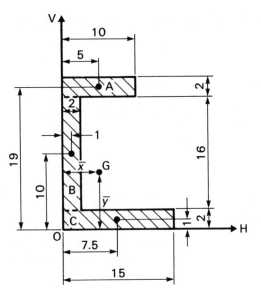

Area	B	D	$A = BD$	x	Ax	Ax^2	$I_{YY}^G = \dfrac{DB^3}{12}$	$I_{OV} = I_{YY}^G + Ax^2$	y	Ay	Ay^2	$I_{XX}^G = \dfrac{BD^3}{12}$	$I_{OH} = I_{XX}^G + Ay^2$
	cm	cm	cm^2	cm	cm^3	cm^4	cm^4	cm^4	cm	cm^3	cm^4	cm^4	cm^4
A	10	2	20	5	100	500	167	667	19	380	7220	7	7227
B	2	16	32	1	32	32	11	43	10	320	3200	683	3883
C	15	2	30	7.5	225	1688	563	2251	1	30	30	10	40
Totals:			82 ΣA		357 ΣAx			2961 ΣI_{OV}		730 ΣAy			11 150 ΣI_{OH}

$$\bar{x} = \frac{\Sigma Ax}{\Sigma A} = \frac{357}{82} = 4.35 \text{ cm} \qquad \bar{y} = \frac{\Sigma Ay}{\Sigma A} = \frac{730}{82} = 8.90 \text{ cm}$$

$$I_{YY} = \Sigma I_{OV} - (\Sigma A)(\bar{x})^2 \qquad\qquad I_{XX} = \Sigma I_{OH} - (\Sigma A)(\bar{y})^2$$

$$= 2961 - (82)(4.35)^2 \qquad\qquad = 11\,150 - (82)(8.90)^2$$

$$= 1410 \text{ cm}^4 \qquad\qquad\qquad = 4650 \text{ cm}^4$$

Example 18.8

Find I_{XX} and I_{YY} for the area shown in Fig. 18.15. All dimensions are in cm.

Fig. **18.15**

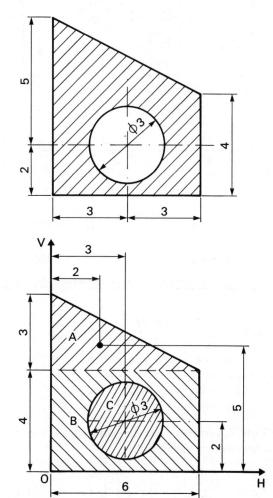

Fig. **18.16**

In the table we shall use for:

a rectangle $\quad I_{YY}^G = \dfrac{DB^3}{12} \quad$ and $\quad I_{XX}^G = \dfrac{BD^3}{12}$

a triangle $\quad I_{YY}^G = \dfrac{DB^3}{36} \quad$ and $\quad I_{XX}^G = \dfrac{BD^3}{36}$

a circle $\quad I_{YY}^G = I_{XX}^G = \dfrac{\pi D^4}{64} \quad$ which is taken as negative, since it is a 'missing' area.

Area	B cm	D cm	A cm²	x cm	Ax cm³	Ax² cm⁴	I_{YY}^G cm⁴	$I_{OV} = I_{YY}^G + Ax^2$ cm⁴	y cm	Ay cm³	Ay² cm⁴	I_{XX}^G cm⁴	$I_{OH} = I_{XX}^G + Ay^2$ cm⁴
A	6	3	9	2	18	36	18	54	5	45	225	5	230
B	6	4	24	3	72	216	72	288	2	48	96	32	128
C	3 diam.		−7.1	3	−21.3	−64	−4	−68	2	−14.2	−28	−4	−32
Totals:			25.9 ΣA		68.7 ΣAx			274 ΣI_{OV}		78.8 ΣAy			326 ΣI_{OH}

$$\bar{x} = \frac{\Sigma Ax}{\Sigma A} = \frac{68.7}{25.9} = 2.65 \text{ cm} \qquad \bar{y} = \frac{\Sigma Ay}{\Sigma A} = \frac{78.8}{25.9} = 3.04 \text{ cm}$$

$$I_{YY} = \Sigma I_{OV} - (\Sigma A)(\bar{x})^2 \qquad I_{XX} = \Sigma I_{OH} - (\Sigma A)(\bar{y})^2$$

$$= 274 - (25.9)(2.65)^2 \qquad\qquad = 326 - (25.9)(3.04)^2$$

$$= 92.1 \text{ cm}^4 \qquad\qquad\qquad = 86.6 \text{ cm}^4$$

EXERCISE

Find I_{XX} and I_{YY} for the following cross-sectional areas. Dimensions are in cm.

1

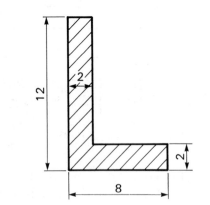

2

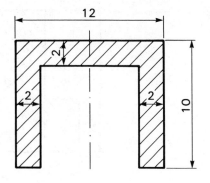

3

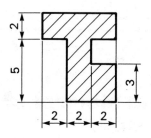

4

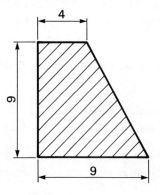

5

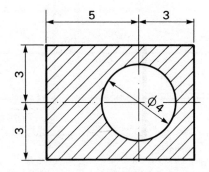

6

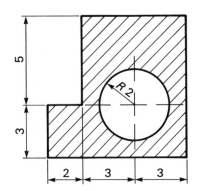

Polar Second Moment of Area

The polar second moment of area, denoted by the symbol J, is the second moment of a circular area about the polar axis (that is the axis passing through the centre of the area and perpendicular to the plane of the area).

One example of its use is in finding stresses due to torsion in a circular shaft.

Now the second moment of area of the elementary strip area, shown in Fig. 18.17, about the reference line CD is Ax^2.

Similarly we say that the polar second moment of area of the elementary circular strip area, shown in Fig. 18.18, is Ar^2.

Fig. **18.17**   Fig. **18.18**

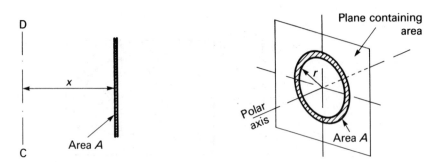

Example 18.9

Find the polar second moment of a circular area of diameter D.

Fig. 18.19 shows the given area and an elementary circular strip. We shall sum the Ar^2 for each strip area to find the polar second moment of area for the whole area.

Fig. **18.19**

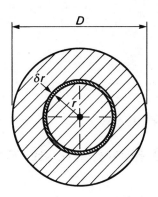

Hence J for the circular area $= \Sigma Ar^2$

Now the approximate area A of the elementary circular strip is

Circular area $=$ Circumference of strip $\times$ Width of strip

$$= 2\pi r \times \delta r$$

$$\therefore \qquad J = \sum_{r=0}^{r=D/2} (2\pi r\, \delta r)r^2$$

$$= \int_0^{D/2} 2\pi r\, dr\, r^2$$

$$= 2\pi \int_0^{D/2} r^3\, dr$$

$$= 2\pi \left[\frac{r^4}{4} \right]_0^{D/2}$$

$$= \frac{\pi D^4}{32}$$

Example 18.10

Find the polar second moment of area of the cross-section of a tube 6 cm outside diameter and 4 cm inside diameter.

We shall consider J of the cross-sectional area shown in Fig. 18.20 to be J of a 6 cm diameter circular area less J of a 4 cm diameter circular area.

$$\therefore \quad \text{required } J = \frac{\pi D^4}{32} - \frac{\pi d^4}{32}$$

$$= \frac{\pi 6^4}{32} - \frac{\pi 4^4}{32}$$

$$= 127 - 25$$

$$= 102 \text{ cm}^4$$

Fig. **18.20**

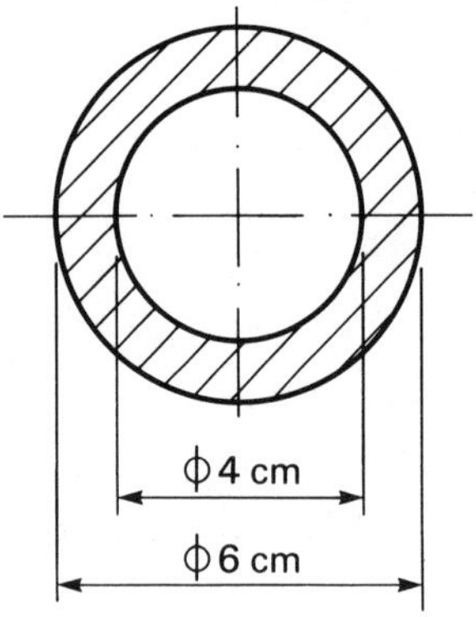

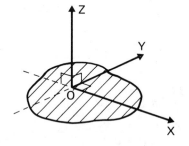

The Perpendicular Axis Theorem

Fig. 18.21 shows a plane area on which the axes OX and OY are drawn, the axes being at right angles. OZ is an axis perpendicular to the plane area. OX, OY and OZ are said to be mutually perpendicular.

Fig. **18.21**

The perpendicular axis theorem states:

$$\boxed{I_{\text{OZ}} = I_{\text{OX}} + I_{\text{OY}}}$$

Example 18.11

Find I_{OZ} for the rectangle shown in Fig. 18.22.

Fig. **18.22**

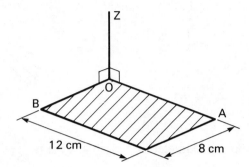

We know that for a rectangle $I_{\text{base}} = \dfrac{bd^3}{3}$

$\therefore \quad I_{OA} = \dfrac{12 \times 8^3}{3} = 2048 \text{ cm}^4 \quad \text{and} \quad I_{OB} = \dfrac{8 \times 12^3}{3} = 4608 \text{ cm}^4$

The perpendicular axis theorem gives

$$I_{OZ} = I_{OA} + I_{OB}$$
$$= 2048 + 4608$$
$$= 6656 \text{ cm}^4$$

Example 18.12

Using the fact that J for a circular area is $\dfrac{\pi d^4}{32}$ and the perpendicular axis theorem find I about a diameter.

Now for a circular area I_{OZ} is called J (Fig. 18.23), and both I_{OX} and I_{OY} are equal to I_{diameter}.

The perpendicular axis theorem states:

$$I_{OZ} = I_{OX} + I_{OY}$$

Fig. **18.23**

$$\therefore \qquad J = I_{\text{diameter}} + I_{\text{diameter}}$$

Thus $\quad I_{\text{diameter}} = \tfrac{1}{2}J$

$$= \frac{1}{2} \times \frac{\pi d^4}{32}$$

$$= \frac{\pi d^4}{64}$$

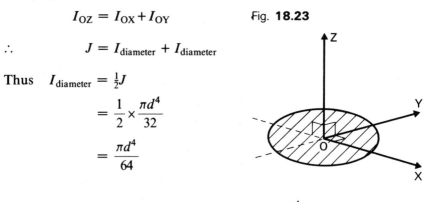

Hence I about a diameter of a circular area is $\dfrac{\pi d^4}{64}$, which confirms a result obtained previously.

A *semicircular area* has I about the boundary diameter equal to one half of the result obtained for a full circular area about a diameter.

Thus for a semicircular area $\quad I_{\text{diameter}} = \dfrac{\pi d^4}{128}$

Example 18.13

The cross-section of a shaft is shown in Fig. 18.24. Find I_{XX}, I_{YY}, and the polar second moment of area J.

Fig. **18.24**

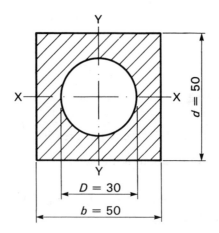

To find I_{XX}

I_{XX} of section $= I_{XX}$ of square $- I_{XX}$ of circle

$$= \frac{bd^3}{12} - \frac{\pi D^4}{64}$$

$$= \frac{50 \times 50^3}{12} - \frac{\pi 30^4}{64}$$

$$= 520\,800 - 39\,800$$

$$= 481\,000 \text{ mm}^4$$

$$= \frac{481\,000}{10^4} \text{ cm}^4 = 48.1 \text{ cm}^4$$

To find I_{YY}

We may see from the symmetry of the figure that $I_{YY} = I_{XX}$

Hence $\qquad\qquad I_{YY} = 48.1 \text{ cm}^4$

To find J

Using the perpendicular axis theorem we have:

$$J = I_{XX} + I_{YY}$$

$$= 48.1 + 48.1$$

$$= 96.2 \text{ cm}^4$$

Radius of Gyration

The formula for finding centroids is

$$(\Sigma A)\bar{x} = \Sigma Ax$$

Similarly for second moments of area we have

$$I = (\Sigma A)k^2 = \Sigma Ax^2$$

where k is called the **radius of gyration** of the area.

In this chapter we have concentrated on using I found from Ax^2. However, we are given sometimes the area, ΣA, of a cross-section together with the radius of gyration k and we then find the second moment of area I by using $I = (\Sigma A)k^2$.

Example 18.14

Find I_{XX} for a cross-sectional area given that the area of the section is
85 cm and $k_{XX} = 3.7$ cm.

We have $\qquad\qquad\qquad I = (\text{area of section})\, k^2$

$\therefore$ $\qquad\qquad\qquad\qquad I_{XX} = 85(3.7)^2$

$\qquad\qquad\qquad\qquad\qquad = 1164 \text{ cm}^4$

EXERCISE

18·3

1 For the cross-sectional area shown in Fig. 18.25 find:

(a) I about a diameter   (b) J about the polar axis   (c) I_{AA}

Fig. **18.25**

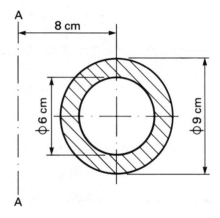

2 For the cross-sectional area shown in Fig. 18.26 find:

(a) I_{XX}, I_{YY} and J (b) k_{XX}

Fig. **18.26**

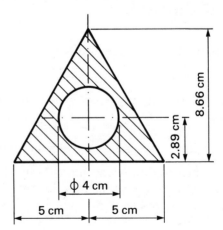

3 Starting with the knowledge that the second moment of area of a rectangle about an edge is $bd^3/3$ and that the second moment of area of a circle about a diameter is $\pi d^4/64$, and also using the perpendicular and parallel axis theorems, find the second moment of the area shown in Fig. 18.27 about the:

(a) axis A'A
(b) axis through G, perpendicular to the plane of the area
(c) axis B'B

Fig. **18.27**

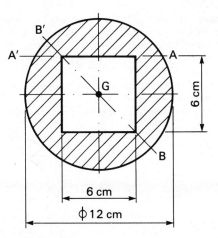

19

Moments of Inertia

OBJECTIVES

1 Determine that the moment of inertia $I = \Sigma mr^2$ for a system of concentrated rotating masses.

2 Define radius of gyration k from the expression $I = (\Sigma m)k^2$.

3 Deduce units of moment of inertia.

4 Understand that all the mass in an elementary ring may be considered to be concentrated at one radius — hence its polar $I = Mr^2$.

5 Use the above information to find by integration the polar I of a solid cylinder.

6 State the parallel axis theorem.

7 Apply the above to typical engineering component shapes.

Moment of Inertia

Moment of inertia is the property of a body used in rotational problems just as the mass is used in problems involving linear motion.

The symbol for moment of inertia is I and must always be stated together with the reference axis about which it has been calculated, in a similar manner as was used for second moment of area. It is unfortunate that engineers have chosen the same symbol for moment of inertia as for second moment of area but, in practice, confusion hardly ever occurs.

The linear kinetic energy of a mass m moving with a velocity v is given by $\frac{1}{2}mv^2$.

Also, the tangential velocity v at the end of a radius r rotating about an axis through O, and perpendicular to the plane of the paper, with angular velocity ω (Fig. 19.1) is given by $v = r\omega$.

Fig. **19.1**

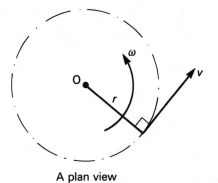

A plan view

Consider the system shown in Fig. 19.2 which comprises three concentrated masses m_1, m_2, and m_3, each being fixed to the end of a radius arm. The radius arms are all attached to a vertical spindle, the arms and spindle being assumed to have negligible mass. The whole system rotates with an angular velocity ω.

Fig. **19.2**

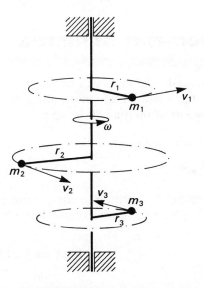

The kinetic energy of the system $= \frac{1}{2}m_1v_1^2 + \frac{1}{2}m_2v_2^2 + \frac{1}{2}m_3v_3^2$

$$= \frac{1}{2}m_1(r_1\omega)^2 + \frac{1}{2}m_2(r_2\omega)^2 + \frac{1}{2}m_3(r_3\omega)^2$$

$$= \frac{1}{2}(m_1r_1^2 + m_2r_2^2 + m_3r_3^2)\omega^2$$

$$= \frac{1}{2}(\Sigma\, mr^2)\omega^2$$

$$= \frac{1}{2}I\omega^2$$

where the moment of inertia of the system is $I = \Sigma\, mr^2$.

Although the expression $I = \Sigma mr^2$ has been derived for three rotating masses it is true for a system comprising any number of concentrated masses.

This expression for I may be used, together with integration, for finding the moment of inertia of bodies whose mass is *not* concentrated at any one particular radius. Example 19.3 shows this method.

Now for first moments of area $\qquad (\Sigma A)\bar{x} = (\Sigma Ax)$

and for second moments of area $\qquad I = (\Sigma A)k^2 = (\Sigma Ax^2)$

Similarly for moments of inertia: $\qquad \boxed{I = (\Sigma m)k^2 = (\Sigma mr^2)}$

where $\qquad\qquad \Sigma m$ is the total mass of the body

and $\qquad\qquad\qquad k$ is the radius of gyration

Units of Moments of Inertia

Since $I = (\Sigma m)k^2$ the units will be those of mass $\times$ distance^2. In basic SI units the mass will be kilograms and the distance metres, and so the units of moment of inertia will be kg m^2.

Example 19.1

Find the moment of inertia of a component about an axis, given that its mass is 5 kg and the radius of gyration about this axis is 200 mm.

Now $\qquad\qquad\qquad I = \text{Total mass} \times k^2$

$$= 5 \times \left(\frac{200}{1000} \right)^2$$

$$= 0.2 \text{ kg m}^2$$

Example 19.2

Find the polar moment of inertia of a rim type flywheel whose dimensions are as shown in Fig. 19.3. The density of the steel from which it is made is 7800 kg/m^3.

Fig. **19.3**

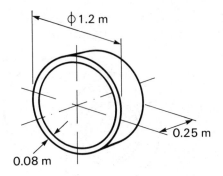

ϕ 1.2 m

0.25 m

0.08 m

A rim type flywheel is a wheel in which all of its mass may be considered to be concentrated round the rim, the hub and spokes being neglected in calculations.

Hence all the mass may be considered to be situated at one particular radius, namely the mean radius of the rim and this may be taken as the radius of gyration.

Hence   Polar $k = 0.6 - 0.04 = 0.56$

   Volume of the rim $= 2\pi \times$ Mean radius $\times$ Width $\times$ Thickness

                $= 2\pi \times 0.56 \times 0.25 \times 0.08$

                $= 0.0704 \text{ m}^3$

and   Mass of the rim $=$ Volume $\times$ Density

                $= 0.0704 \times 7800$

                $= 549 \text{ kg}$

Hence   Polar $I =$ Total mass $\times k^2$

                $= 549 \times 0.56^2$

                $= 172 \text{ kg m}^2$

Example 19.3

Find the moment of inertia, about its polar axis, of a solid cylinder of mass M and radius R.

The mass of a solid cylinder is *not* concentrated at a particular radius and so we shall consider that the cylinder is made up of a series of elementary rings (each similar to the rim of a rim type flywheel) as in Fig. 19.4.

Fig. **19.4**

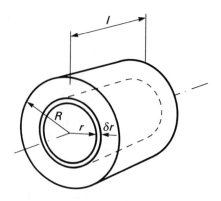

Let the density of the material be ρ, and the length of the cylinder be l.

Now for the cylinder $I = \Sigma\,(mr^2)$

$$= \Sigma\,(\text{mass of elementary ring} \times r^2)$$

$$= \Sigma\,(\text{volume of elementary ring} \times \text{density} \times r^2)$$

$$= \Sigma\,(2\pi r\,\delta r\,l \times p \times r^2)$$

$$= \Sigma\,(2\pi \rho l r^3\,\delta r)$$

$$= \int_0^R 2\pi \rho l r^3\,dr$$

$$= 2\pi \rho l \int_0^R r^3\,dr$$

$$= 2\pi \rho l \left[\frac{r^4}{4}\right]_0^R$$

$$= \frac{2\pi \rho l R^4}{4}$$

$$= \tfrac{1}{2}(\pi R^2 l \rho)R^2$$

$$= \tfrac{1}{2}M R^2$$

since the mass of the cylinder is $M = \pi R^2 l \rho$.

If we also wish to find the radius of gyration k about the polar axis we may use

$$I = Mk^2 \quad \text{since} \quad M = \Sigma m$$

$$\therefore \qquad \tfrac{1}{2}MR^2 = Mk^2$$

from which

$$k^2 = \frac{R^2}{2}$$

or

$$k = \frac{R}{\sqrt{2}}$$

Parallel Axis Theorem

This is similar to that for second moments of area.

In Fig. 19.5 the axis CD is parallel to the axis through the centre of gravity G, h being the distance between the axes. Then:

$$\boxed{I_{CD} = I_{\text{axis through G}} + Mh^2}$$

Fig. **19.5**

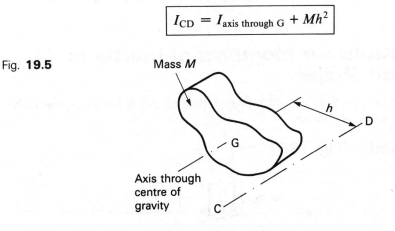

Mass M

Axis through centre of gravity

Example 19.4

Find the moment of inertia for the connecting rod shown in Fig. 19.6, about an axis through its centre of gravity G and parallel to the knife edge. Its mass is 41 kg and I about the knife edge is 15.5 kg m^2 found by oscillating the connecting-rod as a compound pendulum.

Fig. **19.6**

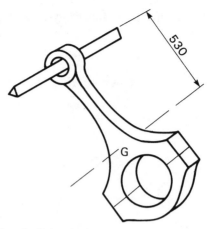

Using the parallel axis theorem,

$$I_{\text{knife edge}} = I_{\text{axis through G}} + Mh^2$$

$\therefore$ $$15.5 = I_{\text{axis through G}} + 41\left(\frac{530}{1000}\right)^2$$

$\therefore$ $$15.5 = I_{\text{axis through G}} + 11.5$$

$\therefore$ $$I_{\text{axis through G}} = 15.5 - 11.5$$

$$= 4 \text{ kg m}^2$$

Expressions for Moments of Inertia for Common Shapes

These may be found in most engineering reference books, typical data being given as follows:

Solid Cylinder (Fig. 19.7)

Fig. **19.7**

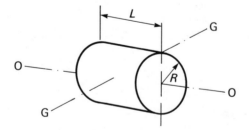

Polar $I_{OO} = \frac{1}{2}MR^2$ and $I_{GG} = M\left(\frac{L^2}{12} + \frac{R^2}{4}\right)$

where GG passes through the centre of gravity and is perpendicular to OO.

Sold Rectangular Block (Fig. 19.8)

Fig. **19.8**

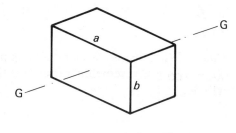

$$I_{GG} = M\left(\frac{a^2 + b^2}{12}\right)$$

where GG passes through the centre of gravity and is perpendicular to face *ab*.

EXERCISE

19·1

1 Find the polar moment of inertia of a solid aluminium cylinder 100 mm diameter and 500 mm long, if the density of aluminium is 2700 kg/m³.

2 Find the radius of gyration k_{GG} of the solid rectangular block shown in Fig. 19.9 if axis GG is perpendicular to the face measuring 200 mm × 300 mm and passes through the centre of gravity of the block.

Fig. **19.9**

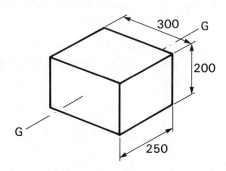

3 Find the polar moment of inertia of a hollow steel cylinder 750 mm long, having an outside diameter of 500 mm and an inside diameter of 400 mm. The density of steel is 7900 kg/m³.

(*Hint:*   The polar moment of inertia of the hollow cylinder is the difference between the polar moments of inertia of a solid cylinder 500 mm in diameter and a solid cylinder 400 mm in diameter.)

4 A hollow copper cylinder is as shown in Fig. 19.10. If the density of copper is 9000 kg/m³ find:

(a) I_{OO}
(b) I_{AA}

(*Hint:*   When using the parallel axis theorem which gives $I_{AA} = I_{OO} + Mh^2$ remember that M is the mass of the hollow cylinder.)

Fig. **19.10**

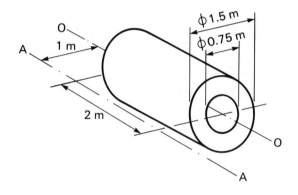

5 Fig. 19.11 shows a solid copper cylinder which has a density of 9000 kg/m³. Find:

(a) I_{GG}, where GG is an axis passing through the centre of gravity and perpendicular to the polar axis
(b) I_{CC}

Fig. **19.11**

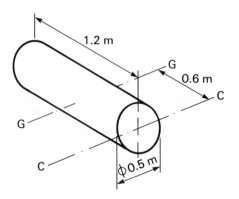

6 Find I_{BB} for the solid block shown in Fig. 19.12 if the density of the material from which it is made is 8000 kg/m³.

Fig. **19.12**

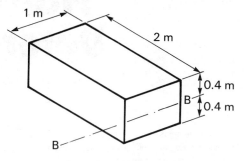

7 The component shown in Fig. 19.13 is made from a material which has a density of 7500 kg/m³. Find I_{DD}.

Fig. **19.13**

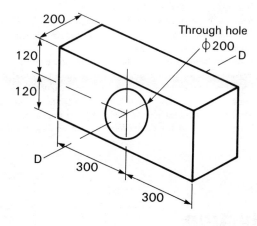

20 Matrices and Determinants

OBJECTIVES

1 Recognise the notation of a matrix.
2 Calculate the sum and difference of two matrices (2×2 only).
3 Calculate the product of two 2×2 matrices.
4 Demonstrate that the product of two matrices is, in general, non-commutative.
5 Define the unit matrix.
6 Recognise the notation for a determinant.
7 Evaluate a 2×2 determinant.
8 Solve simultaneous linear equations with two unknowns using determinants.
9 Describe the meaning of a determinant whose value is zero and define a singular matrix.
10 Obtain the inverse of a 2×2 matrix.
11 Solve simultaneous linear equations with two unknowns by means of matrices.
12 Evaluate the 3×3 determinant by expansion about any row or column.
13 Verify the effect on the sign of a determinant by row/column interchange.
14 Deduce that the value of a determinant is zero if two rows/columns are equal.
15 Verify the effect of extracting a common factor from a row or column.
16 Verify that a determinant is unaltered by addition or subtraction of rows or columns.

Introduction

The block within which a printer sets his type, and a car radiator, could each be called a **matrix**.

A matrix in mathematics is any rectangular array of numbers, usually enclosed in brackets.

Before using matrices in practical problems we first look at how they are ordered, added, subtracted and manipulated in other ways.

Element

Each number or symbol in a matrix is called an **element** of the matrix.

Order

The **dimension** or **order** of a matrix is stated by the number of rows followed by the number of columns in the rectangular array.

e.g.

Matrix	$\begin{pmatrix} 1 & 2 \\ 3 & 4 \end{pmatrix}$	$\begin{pmatrix} a & 2 & -3 \\ 4 & b & x \end{pmatrix}$	$\begin{pmatrix} \sin\theta & 1 \\ \cos\theta & 2 \\ \tan\theta & 3 \end{pmatrix}$	(6)
Order	2×2	2×3	3×2	1×1

Equality

If two matrices are equal, then they must be of the same order and their corresponding elements must be equal.

Thus if $\begin{pmatrix} 2 & 3 & x \\ a & 5 & -2 \end{pmatrix} = \begin{pmatrix} 2 & 3 & 4 \\ -1 & 5 & -2 \end{pmatrix}$ then $x = 4$ and $a = -1$

Addition and Subtraction

Two matrices may be added or subtracted only if they are of the **same order**. We say the matrices are **conformable** for addition (or subtraction) and we add (or subtract) by combining corresponding elements.

Example 20.1

If $A = \begin{pmatrix} 3 & 4 \\ 5 & 6 \end{pmatrix}$ and $B = \begin{pmatrix} 0 & 6 \\ 5 & 2 \end{pmatrix}$ determine:

(a) $C = A + B$
(b) $D = A - B$

(a) $C = \begin{pmatrix} 3 & 4 \\ 5 & 6 \end{pmatrix} + \begin{pmatrix} 0 & 6 \\ 5 & 2 \end{pmatrix} = \begin{pmatrix} 3+0 & 4+6 \\ 5+5 & 6+2 \end{pmatrix} = \begin{pmatrix} 3 & 10 \\ 10 & 8 \end{pmatrix}$

(b) $D = \begin{pmatrix} 3 & 4 \\ 5 & 6 \end{pmatrix} - \begin{pmatrix} 0 & 6 \\ 5 & 2 \end{pmatrix} = \begin{pmatrix} 3-0 & 4-6 \\ 5-5 & 6-2 \end{pmatrix} = \begin{pmatrix} 3 & -2 \\ 0 & 4 \end{pmatrix}$

Zero or Null Matrix

A **zero** or **null** matrix, denoted by **O**, is one in which all the elements are zero. It may be of any order.

Thus $\begin{pmatrix} 0 & 0 \\ 0 & 0 \end{pmatrix}$ is a zero matrix of order 2. It behaves like zero in the real number system.

Identity or Unit Matrix

The **identity** matrix can be of any suitable order with all the main diagonal elements 1 and the remaining elements 0. It is denoted by **I** and behaves like unity in the real number system.

Thus $\begin{pmatrix} 1 & 0 \\ 0 & 1 \end{pmatrix}$ is a unit matrix of order 2.

Transpose

The **transpose** of a matrix **A** is written as **A'** or $\mathbf{A}^T$. When the row of a matrix is interchanged with its corresponding column, that is row 1 becomes column 1, and row 2 becomes column 2 and so on, then the matrix is transposed.

Thus if $\mathbf{A} = \begin{pmatrix} 1 & 2 & -3 \\ 4 & 7 & 0 \end{pmatrix}$ then $\mathbf{A}' = \begin{pmatrix} 1 & 4 \\ 2 & 7 \\ -3 & 0 \end{pmatrix}$

EXERCISE

20·1

1 State the order of each of the following matrices:

(a) $\begin{pmatrix} 1 & 2 \\ 3 & 4 \end{pmatrix}$ (b) $\begin{pmatrix} 5 \\ -6 \end{pmatrix}$ (c) $\begin{pmatrix} a & b & 4 \\ 2 & 3 & 5 \\ x & -6 & 0 \end{pmatrix}$

(d) $\begin{pmatrix} 1 & -2 & -3 & -4 \\ 6 & 2 & 0 & -1 \end{pmatrix}$

2 How many elements are there in:

   (a) a 3×3 matrix   (b) a 2×2 matrix

   (c) a square matrix of order n?

3 Write down the transpose of each matrix in Question 1.

4 Combine the following matrices:

   (a) $\begin{pmatrix} 2 & 1 \\ 3 & 2 \end{pmatrix} + \begin{pmatrix} -2 & -1 \\ 6 & 0 \end{pmatrix}$ (b) $\begin{pmatrix} 2 & 1 \\ 3 & 2 \end{pmatrix} - \begin{pmatrix} -2 & -1 \\ 6 & 0 \end{pmatrix}$

   (c) $\begin{pmatrix} \frac{1}{2} & 1 \\ \frac{1}{3} & \frac{1}{5} \end{pmatrix} + \begin{pmatrix} \frac{1}{3} & -\frac{1}{2} \\ \frac{1}{2} & \frac{4}{5} \end{pmatrix}$

5 Find a, b and c if $(a \quad b \quad c) - (-3 \quad 4 \quad 1) = (-5 \quad 1 \quad 0)$.

6 Complete $\begin{pmatrix} \frac{1}{2} & \frac{1}{4} \\ \frac{1}{5} & \frac{1}{6} \end{pmatrix} - \begin{pmatrix} \frac{1}{6} & \frac{1}{5} \\ \frac{1}{6} & \frac{1}{9} \end{pmatrix}$.

7 Solve the equation $X - \begin{pmatrix} 1 & 3 \\ 5 & -2 \end{pmatrix} = \begin{pmatrix} 4 & 5 \\ 7 & 0 \end{pmatrix}$ where X is a 2×2 matrix.

8 If $\begin{pmatrix} 4 \\ 5 \end{pmatrix} + \begin{pmatrix} x \\ y \end{pmatrix} = \begin{pmatrix} 4 \\ 10 \end{pmatrix}$,   determine $\begin{pmatrix} x \\ y \end{pmatrix}$.

Multiplication of a Matrix by a Real Number

A matrix may be multiplied by a number in the following way:

$$4\begin{pmatrix} 2 & 3 \\ 7 & -1 \end{pmatrix} = \begin{pmatrix} 4 \times 2 & 4 \times 3 \\ 4 \times 7 & 4 \times (-1) \end{pmatrix} = \begin{pmatrix} 8 & 12 \\ 28 & -4 \end{pmatrix}$$

Conversely the common factor of each element in a matrix may be written outside the matrix. Thus $\begin{pmatrix} 9 & 3 \\ 42 & 15 \end{pmatrix} = 3\begin{pmatrix} 3 & 1 \\ 14 & 5 \end{pmatrix}$.

Matrix Multiplication

Two matrices can only be multiplied together if the number of columns in the first matrix is equal to the number of rows in the second matrix. We say that the matrices are **conformable** for multiplication. The method for multiplying together a pair of 2×2 matrices is as follows:

$$\begin{pmatrix} a & b \\ c & d \end{pmatrix} \times \begin{pmatrix} e & f \\ g & h \end{pmatrix} = \begin{pmatrix} ae + bg & af + bh \\ ce + dg & cf + dh \end{pmatrix}$$

Example 20.2

(a) $\begin{pmatrix} 2 & 3 \\ 4 & 5 \end{pmatrix} \times \begin{pmatrix} 7 & 1 \\ 0 & 6 \end{pmatrix} = \begin{pmatrix} (2 \times 7) + (3 \times 0) & (2 \times 1) + (3 \times 6) \\ (4 \times 7) + (5 \times 0) & (4 \times 1) + (5 \times 6) \end{pmatrix}$

$$= \begin{pmatrix} 14 & 20 \\ 28 & 34 \end{pmatrix}$$

(b) $\begin{pmatrix} 3 & 4 \\ 2 & 5 \end{pmatrix} \times \begin{pmatrix} 6 \\ 7 \end{pmatrix} = \begin{pmatrix} (3 \times 6) + (4 \times 7) \\ (2 \times 6) + (5 \times 7) \end{pmatrix} = \begin{pmatrix} 46 \\ 47 \end{pmatrix}$

(c) $\begin{pmatrix} 3 \\ 2 \end{pmatrix} \times \begin{pmatrix} 4 & 6 \\ 5 & 7 \end{pmatrix}$ This is not possible since the matrices are not conformable.

Example 20.3

Form the products **AB** and **BA** given that $\mathbf{A} = \begin{pmatrix} 1 & 2 \\ 3 & 4 \end{pmatrix}$ and $\mathbf{B} = \begin{pmatrix} 5 & 6 \\ 7 & 8 \end{pmatrix}$ and hence show that $\mathbf{AB} \neq \mathbf{BA}$.

$\mathbf{AB} = \begin{pmatrix} 1 & 2 \\ 3 & 4 \end{pmatrix}\begin{pmatrix} 5 & 6 \\ 7 & 8 \end{pmatrix} = \begin{pmatrix} (1 \times 5) + (2 \times 7) & (1 \times 6) + (2 \times 8) \\ (3 \times 5) + (4 \times 7) & (3 \times 6) + (4 \times 8) \end{pmatrix}$

$$= \begin{pmatrix} 19 & 22 \\ 43 & 50 \end{pmatrix}$$

$\mathbf{BA} = \begin{pmatrix} 5 & 6 \\ 7 & 8 \end{pmatrix}\begin{pmatrix} 1 & 2 \\ 3 & 4 \end{pmatrix} = \begin{pmatrix} (5 \times 1) + (6 \times 3) & (5 \times 2) + (6 \times 4) \\ (7 \times 1) + (8 \times 3) & (7 \times 2) + (8 \times 4) \end{pmatrix}$

$$= \begin{pmatrix} 23 & 34 \\ 31 & 46 \end{pmatrix}$$

As we see the results are different and, in general, *matrix multiplication is non-commutative,* i.e. $\mathbf{AB} \neq \mathbf{BA}$.

EXERCISE

20·2   **1** If $A = \begin{pmatrix} 3 & 0 \\ -2 & 1 \end{pmatrix}$ and $B = \begin{pmatrix} -4 & 1 \\ 3 & -2 \end{pmatrix}$ determine:

   (a) $2A$ (b) $3B$ (c) $2A + 3B$ (d) $2A - 3B$

2 Calculate the following products:

   (a) $\begin{pmatrix} 3 & 1 \\ 2 & 0 \end{pmatrix}\begin{pmatrix} 4 & -1 \\ 2 & 3 \end{pmatrix}$ (b) $\begin{pmatrix} 2 & 1 \\ 3 & 1 \end{pmatrix}\begin{pmatrix} 1 & 0 \\ 0 & 1 \end{pmatrix}$

   (c) $\begin{pmatrix} 2 & 1 \\ 4 & 2 \end{pmatrix}\begin{pmatrix} 2 & 3 \\ 1 & 5 \end{pmatrix}$ (d) $\begin{pmatrix} 1 & 0 \\ 0 & 1 \end{pmatrix}\begin{pmatrix} a & b \\ c & d \end{pmatrix}$

   (e) $\begin{pmatrix} k & 0 \\ 0 & k \end{pmatrix}\begin{pmatrix} a & b \\ c & d \end{pmatrix}$

3 If $A = \begin{pmatrix} 1 & 2 \\ 3 & 4 \end{pmatrix}$ and $B = \begin{pmatrix} 2 & -1 \\ 1 & 3 \end{pmatrix}$ calculate:

   (a) A^2 (that is $A \times A$)   (b) B^2 (c) $2AB$
   (d) $A^2 + B^2 + 2AB$ (e) $(A + B)^2$

Determinant of a Square Matrix of Order 2

If matrix $A = \begin{pmatrix} a & b \\ c & d \end{pmatrix}$ then its **determinant** is denoted by $|A|$ or det A
and the result is a **number** given by

$$|A| = \begin{vmatrix} a & b \\ c & d \end{vmatrix} = ad - bc$$

Example 20.4

Evaluate $|A|$ if $A = \begin{pmatrix} 1 & -2 \\ 3 & 4 \end{pmatrix}$.

$$|A| = \begin{vmatrix} 1 & -2 \\ 3 & 4 \end{vmatrix} = 1 \times 4 - (-2) \times 3 = 10$$

Solution of Simultaneous Linear Equations Using Determinants

To solve simultaneous linear equations with two unknowns using determinants, the following procedure is used.

(1) Write out the two equations in order: $a_1x + b_1y = c_1$

$$a_2x + b_2y = c_2$$

(2) Calculate $\Delta = \begin{vmatrix} a_1 & b_1 \\ a_2 & b_2 \end{vmatrix}$

(3) Then $x = \dfrac{\begin{vmatrix} c_1 & b_1 \\ c_2 & b_2 \end{vmatrix}}{\Delta}$ and $y = \dfrac{\begin{vmatrix} a_1 & c_1 \\ a_2 & c_2 \end{vmatrix}}{\Delta}$

Example 20.5

By using determinants, solve the simultaneous equations

$$3x + 4y = 22$$
$$2x + 5y = 24$$

Now $\Delta = \begin{vmatrix} 3 & 4 \\ 2 & 5 \end{vmatrix} = (3 \times 5) - (4 \times 2) = 7$

Thus $x = \dfrac{\begin{vmatrix} 22 & 4 \\ 24 & 5 \end{vmatrix}}{7} = \dfrac{(22 \times 5) - (4 \times 24)}{7} = \dfrac{14}{7} = 2$

And $y = \dfrac{\begin{vmatrix} 3 & 22 \\ 2 & 24 \end{vmatrix}}{7} = \dfrac{(3 \times 24) - (22 \times 2)}{7} = \dfrac{28}{7} = 4$

EXERCISE

20·3   **1** Evaluate the following determinants:

(a) $\begin{vmatrix} 5 & 2 \\ 3 & 6 \end{vmatrix}$ (b) $\begin{vmatrix} 7 & 4 \\ 5 & 2 \end{vmatrix}$.   (c) $\begin{vmatrix} 6 & 8 \\ 2 & 5 \end{vmatrix}$

2 Solve the following simultaneous equations by using determinants:

(a) $3x + 4y = 11$
$x + 7y = 15$

(b) $5x + 3y = 29$
$4x + 7y = 37$

(c) $4x - 6y = -2.5$
$7x - 5y = -0.25$

The Inverse of a Square Matrix of Order 2

Instead of dividing a number by 5 we can multiply by $\frac{1}{5}$ and obtain the same result.

Thus $\frac{1}{5}$ is the multiplicative inverse of 5. That is $5 \times \frac{1}{5} = 1$.

In matrix algebra we never divide by a matrix but multiply instead by the inverse. The inverse of matrix $\mathbf{A}$ is denoted by $\mathbf{A}^{-1}$ and is such that

$$\mathbf{AA}^{-1} = \begin{pmatrix} 1 & 0 \\ 0 & 1 \end{pmatrix} = \mathbf{I}, \quad \text{the identity matrix}$$

To find the inverse $\mathbf{A}^{-1}$ of the square matrix $\mathbf{A} = \begin{pmatrix} a & b \\ c & d \end{pmatrix}$

we use the expression: $\mathbf{A}^{-1} = \dfrac{1}{|\mathbf{A}|} \begin{pmatrix} d & -b \\ -c & a \end{pmatrix} = \dfrac{1}{ad - bc} \begin{pmatrix} d & -b \\ -c & a \end{pmatrix}$

Example 20.6

Determine the inverse of $\mathbf{A} = \begin{pmatrix} 1 & -2 \\ 3 & 4 \end{pmatrix}$ and verify the result.

Now $\qquad |\mathbf{A}| = \begin{vmatrix} 1 & -2 \\ 3 & 4 \end{vmatrix} = (1 \times 4) - (3 \times -2) = 10$

Hence $\qquad \mathbf{A}^{-1} = \frac{1}{10} \begin{pmatrix} 4 & 2 \\ -3 & 1 \end{pmatrix} = \begin{pmatrix} 0.4 & 0.2 \\ -0.3 & 0.1 \end{pmatrix}$

To verify the result we have

$$\mathbf{AA}^{-1} = \begin{pmatrix} 1 & -2 \\ 3 & 4 \end{pmatrix} \begin{pmatrix} 0.4 & 0.2 \\ -0.3 & 0.1 \end{pmatrix} = \begin{pmatrix} 1 & 0 \\ 0 & 1 \end{pmatrix} = \mathbf{I}$$

Singular Matrix

A matrix which does not have an inverse is called a **singular matrix**.
This happens when $|\mathbf{A}| = 0$

For example, since $\begin{vmatrix} 3 & 6 \\ 1 & 2 \end{vmatrix} = (3 \times 2) - (6 \times 1) = 0$

then $\begin{pmatrix} 3 & 6 \\ 1 & 2 \end{pmatrix}$ is a singular matrix.

EXERCISE

20·4

Decide whether each of the matrices in Questions 1–9 has an inverse. If the inverse exists, find it.

1 $\begin{pmatrix} 2 & 5 \\ 1 & 4 \end{pmatrix}$ **2** $\begin{pmatrix} 2 & 5 \\ 1 & 3 \end{pmatrix}$ **3** $\begin{pmatrix} 3 & 2 \\ 1 & 2 \end{pmatrix}$

4 $\begin{pmatrix} 4 & 10 \\ 2 & 5 \end{pmatrix}$ **5** $\begin{pmatrix} 224 & 24 \\ 24 & 4 \end{pmatrix}$ **6** $\begin{pmatrix} a & -b \\ -a & b \end{pmatrix}$

7 $\begin{pmatrix} 2 & 3 \\ -1 & 1 \end{pmatrix}$ **8** $\begin{pmatrix} 2 & -3 \\ 1 & 5 \end{pmatrix}$ **9** $\begin{pmatrix} 1 & 1 \\ 0 & 1 \end{pmatrix}$

10 Given that $\mathbf{A} = \begin{pmatrix} 1 & 0 \\ 3 & 2 \end{pmatrix}$ and $\mathbf{B} = \begin{pmatrix} 3 & 5 \\ 1 & 2 \end{pmatrix}$, calculate:

(a) $\mathbf{A}^{-1}$ (b) $\mathbf{B}^{-1}$ (c) $\mathbf{B}^{-1}\mathbf{A}^{-1}$ (d) $\mathbf{AB}$
(e) $(\mathbf{AB})^{-1}$ (f) Compare the answers to (c) and (e).

Systems of Linear Equations

Given the system of equations $\left. \begin{aligned} 5x + y &= 7 \\ 3x - 4y &= 18 \end{aligned} \right\}$

We can rewrite it in the form $\begin{pmatrix} 5x + y \\ 3x - 4y \end{pmatrix} = \begin{pmatrix} 7 \\ 18 \end{pmatrix}$

or $\begin{pmatrix} 5 & 1 \\ 3 & -4 \end{pmatrix}\begin{pmatrix} x \\ y \end{pmatrix} = \begin{pmatrix} 7 \\ 18 \end{pmatrix}$

That is $\left(\begin{matrix}\text{Matrix of} \\ \text{coefficients}\end{matrix}\right)\left(\begin{matrix}\text{Matrix of} \\ \text{variables}\end{matrix}\right) = \left(\begin{matrix}\text{Matrix of} \\ \text{constants}\end{matrix}\right)$

Denote the matrix of coefficients by $\mathbf{C}$ and its inverse by $\mathbf{C}^{-1}$.

Then $$|\mathbf{C}| = \begin{vmatrix} 5 & 1 \\ 3 & -4 \end{vmatrix} = 5 \times (-4) - 1 \times 3 = -23$$

and $$\mathbf{C}^{-1} = \frac{1}{-23}\begin{pmatrix} -4 & -1 \\ -3 & 5 \end{pmatrix} = \frac{1}{23}\begin{pmatrix} 4 & 1 \\ 3 & -5 \end{pmatrix}$$

Now $$\mathbf{C}\begin{pmatrix} x \\ y \end{pmatrix} = \begin{pmatrix} 7 \\ 18 \end{pmatrix}$$

and multiplying both sides by $\mathbf{C}^{-1}$ gives

$$\mathbf{C}^{-1}\mathbf{C}\begin{pmatrix} x \\ y \end{pmatrix} = \mathbf{C}^{-1}\begin{pmatrix} 7 \\ 18 \end{pmatrix}$$

$\therefore$ $$\mathbf{I}\begin{pmatrix} x \\ y \end{pmatrix} = \mathbf{C}^{-1}\begin{pmatrix} 7 \\ 18 \end{pmatrix}$$

or $$\begin{pmatrix} 1 & 0 \\ 0 & 1 \end{pmatrix} \times \begin{pmatrix} x \\ y \end{pmatrix} = \frac{1}{23}\begin{pmatrix} 4 & 1 \\ 3 & -5 \end{pmatrix} \times \begin{pmatrix} 7 \\ 18 \end{pmatrix}$$

$\therefore$ $$\begin{pmatrix} 1 \times x & 0 \times y \\ 0 \times x & 1 \times y \end{pmatrix} = \frac{1}{23}\begin{pmatrix} 4 \times 7 + & 1 \times 18 \\ 3 \times 7 + (-5) \times 18 \end{pmatrix}$$

$\therefore$ $$\begin{pmatrix} x \\ y \end{pmatrix} = \frac{1}{23}\begin{pmatrix} 46 \\ -69 \end{pmatrix}$$

$\therefore$ $$\begin{pmatrix} x \\ y \end{pmatrix} = \begin{pmatrix} 2 \\ -3 \end{pmatrix}$$

Thus comparing the matrices shows that $x = 2$ and $y = -3$.

We would not normally perform multiplication by the unit matrix.

We did so here to illustrate that when a matrix, here $\begin{pmatrix} x \\ y \end{pmatrix}$, is multiplied by the unit matrix then it is unaltered.

This confirms that the unit matrix performs as unity (the number one) in normal arithmetic.

Use matrix methods to solve each of the following systems of equations:

1 $\left.\begin{array}{r} x + y = 1 \\ 3x + 2y = 8 \end{array}\right\}$ **2** $\left.\begin{array}{r} x + y = 6 \\ 3x - 2y = -7 \end{array}\right\}$ **3** $\left.\begin{array}{r} 5x - 2y = 17 \\ 2x + 3y = 3 \end{array}\right\}$

4 $\left.\begin{array}{r} 3x - 2y = 12 \\ 4x + y = 5 \end{array}\right\}$ **5** $\left.\begin{array}{r} 3x + 2y = 6 \\ 4x - y = 5 \end{array}\right\}$ **6** $\left.\begin{array}{r} 3x - 4y = 26 \\ 5x + 6y = -20 \end{array}\right\}$

Determinants of the Third Order (3 × 3 Determinants)

A determinant of the third order contains 3 rows and 3 columns as, for example:

$$\begin{vmatrix} 1 & 6 & 9 \\ 3 & 5 & 4 \\ 2 & 7 & 8 \end{vmatrix}$$

If we cross out the row and column through element 1,

$$\begin{vmatrix} 1 & 6 & 9 \\ 3 & 5 & 4 \\ 2 & 7 & 8 \end{vmatrix}$$

This leaves the determinant $\begin{vmatrix} 5 & 4 \\ 7 & 8 \end{vmatrix}$ which is called the **minor** of element 1.

Similarly, if we cross out the row and column through element 4, we have

$$\begin{vmatrix} 1 & 6 & 9 \\ 3 & 5 & 4 \\ 2 & 7 & 8 \end{vmatrix}$$

This leaves the determinant $\begin{vmatrix} 1 & 6 \\ 2 & 7 \end{vmatrix}$ which is called the minor of element 4.

To expand a 3×3 determinant we may write the products of each top row element and their respective minors, giving the terms alternately a plus or minus sign. Thus:

$$\begin{vmatrix} 1 & 6 & 9 \\ 3 & 5 & 4 \\ 2 & 7 & 8 \end{vmatrix} = 1\begin{vmatrix} 5 & 4 \\ 7 & 8 \end{vmatrix} - 6\begin{vmatrix} 3 & 4 \\ 2 & 8 \end{vmatrix} + 9\begin{vmatrix} 3 & 5 \\ 2 & 7 \end{vmatrix}$$

$$= 1(5 \times 8 - 4 \times 7) - 6(3 \times 8 - 4 \times 2) + 9(3 \times 7 - 5 \times 2)$$

$$= 1(40 - 28) - 6(24 - 8) + 9(21 - 10)$$

$$= 1(12) - 6(16) + 9(11)$$

$$= 12 - 96 + 99$$

$$= 15$$

We may expand a determinant along any row or column in the same way, provided we give to each product of an element and its minor the + or − sign as shown in the pattern below:

$$
\begin{array}{ccc}
+ & - & + \\
- & + & - \\
+ & - & +
\end{array}
$$

So if we expand about the first column, then

$$
\begin{vmatrix} 1 & 6 & 9 \\ 3 & 5 & 4 \\ 2 & 7 & 8 \end{vmatrix} = 1\begin{vmatrix} 5 & 4 \\ 7 & 8 \end{vmatrix} - 3\begin{vmatrix} 6 & 9 \\ 7 & 8 \end{vmatrix} + 2\begin{vmatrix} 6 & 9 \\ 5 & 4 \end{vmatrix}
$$

$$
= 1(5 \times 8 - 4 \times 7) - 3(6 \times 8 - 9 \times 7) + 2(6 \times 4 - 9 \times 5)
$$

$$
= 1(40 - 28) - 3(48 - 63) + 2(24 - 45)
$$

$$
= 1(12) - 3(-15) + 2(-21)
$$

$$
= 12 + 45 - 42
$$

$$
= 15 \quad \text{which was obtained previously}
$$

And if we expand about the middle row, then

$$
\begin{vmatrix} 1 & 6 & 9 \\ 3 & 5 & 4 \\ 2 & 7 & 8 \end{vmatrix} = -3\begin{vmatrix} 6 & 9 \\ 7 & 8 \end{vmatrix} + 5\begin{vmatrix} 1 & 9 \\ 2 & 8 \end{vmatrix} - 4\begin{vmatrix} 1 & 6 \\ 2 & 7 \end{vmatrix}
$$

$$
= -3(6 \times 8 - 9 \times 7) + 5(1 \times 8 - 9 \times 2) - 4(1 \times 7 - 6 \times 2)
$$

$$
= -3(48 - 63) + 5(8 - 18) - 4(7 - 12)
$$

$$
= -3(-15) + 5(-10) - 4(-5)
$$

$$
= 45 - 50 + 20
$$

$$
= 15 \quad \text{again as obtained previously}
$$

In general it is possible to expand a determinant about any row or column provided we use the pattern for the + and − signs.

General Properties of Determinants

These are really a set of rules, which we will either establish or verify, to help simplify a determinant before we expand it, and thus save work. It is possible to verify all these rules by using determinants with number elements, as we did previously. Although it may not appear so at first sight it is often clearer to use symbols for the elements — the working is not at all complicated and only requires a little patience.

Interchange of Rows or Columns

If we use a determinant with standard notation and expand about the first column, then

$$\begin{vmatrix} a_1 & b_1 & c_1 \\ a_2 & b_2 & c_2 \\ a_3 & b_3 & c_3 \end{vmatrix} = a_1 \begin{vmatrix} b_2 & c_2 \\ b_3 & c_3 \end{vmatrix} - a_2 \begin{vmatrix} b_1 & c_1 \\ b_3 & c_3 \end{vmatrix} + a_3 \begin{vmatrix} b_1 & c_1 \\ b_2 & c_2 \end{vmatrix}$$

Now if we interchange the first and second columns, and again expand about the second column (remembering the pattern for + and − signs), then

$$\begin{vmatrix} b_1 & a_1 & c_1 \\ b_2 & a_2 & c_2 \\ b_3 & a_3 & c_3 \end{vmatrix} = -a_1 \begin{vmatrix} b_2 & c_2 \\ b_3 & c_3 \end{vmatrix} + a_2 \begin{vmatrix} b_1 & c_1 \\ b_3 & c_3 \end{vmatrix} - a_3 \begin{vmatrix} b_1 & c_1 \\ b_2 & c_2 \end{vmatrix}$$

Now if we examine the right-hand sides of these expressions we see that they are the same except that the signs are different. You may like to try this for yourself by first expanding about the top row — then exchange the top and bottom rows and expand about the bottom row. Again the signs will change. This means that the value of the determinant is unaltered but its sign has changed. Thus we have established the rule:

> If two rows (or columns) are interchanged the determinant changes sign.

A Row or Column Multiplied by a Common Factor

Suppose that the elements in the first column are multiplied by 5. Then 5 is called a common factor of the first column. If we expand about the first column, then

$$\begin{vmatrix} 5a_1 & b_1 & c_1 \\ 5a_2 & b_2 & c_2 \\ 5a_3 & b_3 & c_3 \end{vmatrix} = 5a_1 \begin{vmatrix} b_2 & c_2 \\ b_3 & c_3 \end{vmatrix} - 5a_2 \begin{vmatrix} b_1 & c_1 \\ b_3 & c_3 \end{vmatrix} + 5a_3 \begin{vmatrix} b_1 & c_1 \\ b_2 & c_2 \end{vmatrix}$$

By comparing the right-hand side of this expression with that for the original determinant we see that each term has been multiplied by 5. This means that the value of this determinant is 5 times that of the original.

Hence we may deduce the rule:

> If one row (or column) is multiplied by a common factor k, then the resulting determinant will have a value k times that of the original.

Now let us see how this rule helps in evaluating a determinant.

Example 20.7

Evaluate $\Delta = \begin{vmatrix} 36 & 1 & 6 \\ 18 & 3 & -2 \\ -63 & 35 & 21 \end{vmatrix}$

The symbol Δ is often used to represent a determinant just as I is used for an integral.

Now for the working out, and it looks as if we have rather large numbers to handle. However, if we examine the first column we see that there is a common factor of 9. So we may rewrite Δ as follows:

$$\Delta = \begin{vmatrix} 9(4) & 1 & 6 \\ 9(2) & 3 & -2 \\ 9(-7) & 35 & 21 \end{vmatrix} = 9 \begin{vmatrix} 4 & 1 & 6 \\ 2 & 3 & -2 \\ -7 & 35 & 21 \end{vmatrix}$$

Much better now! Another look at the bottom row shows that 7 is a common factor, so we may rewrite Δ again as:

$$\Delta = 9 \begin{vmatrix} 4 & 1 & 6 \\ 2 & 3 & -2 \\ 7(-1) & 7(5) & 7(3) \end{vmatrix} = 9 \times 7 \begin{vmatrix} 4 & 1 & 6 \\ 2 & 3 & -2 \\ -1 & 5 & 3 \end{vmatrix}$$

And we can now easily work out the value of Δ, expanding about the top line:

$$\Delta = 63 \left\{ 4 \begin{vmatrix} 3 & -2 \\ 5 & 3 \end{vmatrix} - 1 \begin{vmatrix} 2 & -2 \\ -1 & 3 \end{vmatrix} + 6 \begin{vmatrix} 2 & 3 \\ -1 & 5 \end{vmatrix} \right\}$$

$$= 63 \{ 4[3 \times 3 - (-2)5] - 1[2 \times 3 - (-2)(-1)] + 6[2 \times 5 - 3(-1)] \}$$

$$= 63 \{ 4[9 + 10] - 1[6 - 2] + 6[10 + 3] \}$$

$$= 63[4(19) - 1(4) + 6(13)]$$

$$= 63[76 - 4 + 78]$$

$$= 9450$$

Addition (or Subtraction) of Rows or Columns

Using a determinant with standard notation and expanding about the first column,

$$\begin{vmatrix} a_1 & b_1 & c_1 \\ a_2 & b_2 & c_2 \\ a_3 & b_3 & c_3 \end{vmatrix} = a_1 \begin{vmatrix} b_2 & c_2 \\ b_3 & c_3 \end{vmatrix} - a_2 \begin{vmatrix} b_1 & c_1 \\ b_3 & c_3 \end{vmatrix} + a_3 \begin{vmatrix} b_1 & c_1 \\ b_2 & c_2 \end{vmatrix}$$

$$= a_1(b_2c_3 - b_3c_2) - a_2(b_1c_3 - b_3c_1) + a_3(b_1c_2 - b_2c_1)$$

Now let us add the elements of the last column to each of the corresponding elements of the centre column, and then expand about the first column:

$$\begin{vmatrix} a_1 & b_1+c_1 & c_1 \\ a_2 & b_2+c_2 & c_2 \\ a_3 & b_3+c_3 & c_3 \end{vmatrix} = a_1 \begin{vmatrix} b_2+c_2 & c_2 \\ b_3+c_3 & c_3 \end{vmatrix} - a_2 \begin{vmatrix} b_1+c_1 & c_1 \\ b_3+c_3 & c_3 \end{vmatrix} + a_3 \begin{vmatrix} b_1+c_1 & c_1 \\ b_2+c_2 & c_2 \end{vmatrix}$$

$$= a_1[(b_2+c_2)c_3 - c_2(b_3+c_3)]$$
$$- a_2[(b_1+c_1)c_3 - c_1(b_3+c_3)]$$
$$+ a_3[(b_1+c_1)c_2 - c_1(b_2+c_2)]$$

$$= a_1[b_2c_3 + \cancel{c_2c_3} - b_3c_2 - \cancel{c_2c_3}]$$
$$- a_2[b_1c_3 + \cancel{c_1c_3} - b_3c_1 - \cancel{c_1c_3}]$$
$$+ a_3[b_1c_2 + \cancel{c_1c_2} - b_2c_1 - \cancel{c_1c_2}]$$

$$= a_1[b_2c_3 - b_3c_2] - a_2[b_1c_3 - b_3c_1] + a_3[b_1c_2 - b_2c_1]$$

This result is the same as that for the original determinant. If we had added three times the elements of the last column (i.e. $3c_1$, $3c_2$ and $3c_3$) to each of the corresponding elements of the centre column we can see, by inspection, that the same terms would have cancelled out and the result would still be the same.

You may like to try this for yourself by subtracting four times the elements of the second row (i.e. $4a_2$, $4b_2$ and $4c_2$) from each of the corresponding elements of the last row, and then expand about the first row. This gives the same result again.

Hence we may deduce the rule:

> The value of a determinant is unaltered if a multiple of any row (or column) is added to any other row (or column).

This is really a labour-saving device, as the following examples show.

Example 20.8

Evaluate $\Delta = \begin{vmatrix} 6 & 3 & 3 \\ 8 & 7 & 2 \\ 5 & 6 & 3 \end{vmatrix}$.

We see immediately that subtracting the last column from the centre column gives

$$\Delta = \begin{vmatrix} 6 & 3-3 & 3 \\ 8 & 7-2 & 2 \\ 5 & 6-3 & 3 \end{vmatrix} = \begin{vmatrix} 6 & 0 & 3 \\ 8 & 5 & 2 \\ 5 & 3 & 3 \end{vmatrix}$$

and if we then subtract twice the last column from the first column,

$$\Delta = \begin{vmatrix} 6-2(3) & 0 & 3 \\ 8-2(2) & 5 & 2 \\ 5-2(3) & 3 & 3 \end{vmatrix} = \begin{vmatrix} 0 & 0 & 3 \\ 4 & 5 & 2 \\ -1 & 3 & 3 \end{vmatrix}$$

and if we now expand about the top row:

$$\Delta = 0 \begin{vmatrix} 5 & 2 \\ 3 & 3 \end{vmatrix} - 0 \begin{vmatrix} 4 & 2 \\ -1 & 3 \end{vmatrix} + 3 \begin{vmatrix} 4 & 5 \\ -1 & 3 \end{vmatrix}$$

$$= 3[4 \times 3 - 5(-1)]$$

$$= 51$$

Example 20.9

Evaluate $\quad \Delta = \begin{vmatrix} 8 & 3 & 4 \\ 7 & 1 & 1 \\ 9 & -2 & 3 \end{vmatrix}$.

How we simplify is entirely up to us — here we may decide to subtract the last column from the centre column to make the centre element zero. But, as shown in the last example, we usually expand about the row or column which contains the zero elements, and this will be either the centre column or middle row. Most readers will soon have a favourite row or column about which they prefer to expand a determinant.

If this happens to be the first column then this sequence will be of interest:

Subtract three times the last column from the first column:

$$\Delta = \begin{vmatrix} 8-3(4) & 3 & 4 \\ 7-3(1) & 1 & 1 \\ 9-3(3) & -2 & 3 \end{vmatrix} = \begin{vmatrix} -4 & 3 & 4 \\ 4 & 1 & 1 \\ 0 & -2 & 3 \end{vmatrix}$$

Now add the top row to the middle row:

$$\Delta = \begin{vmatrix} -4 & 3 & 4 \\ 4+(-4) & 1+3 & 1+4 \\ 0 & -2 & 3 \end{vmatrix} = \begin{vmatrix} -4 & 3 & 4 \\ 0 & 4 & 5 \\ 0 & -2 & 3 \end{vmatrix}$$

And expand about the first column to give:

$$\Delta = -4 \begin{vmatrix} 4 & 5 \\ -2 & 3 \end{vmatrix}$$

$$= -4[4 \times 3 - 5(-2)]$$

$$= -88$$

Determinants with Two Equal Columns (or Rows)

Example 20.10

$$\Delta = \begin{vmatrix} 6 & 6 & 5 \\ 3 & 3 & -1 \\ 2 & 2 & 7 \end{vmatrix}$$

Subtract the centre column from the first column:-

$$\Delta = \begin{vmatrix} 6-6 & 6 & 5 \\ 3-3 & 3 & -1 \\ 2-2 & 2 & 7 \end{vmatrix} = \begin{vmatrix} 0 & 6 & 5 \\ 0 & 3 & -1 \\ 0 & 2 & 7 \end{vmatrix}$$

Now it can be seen that if we expand about the first column the value of the determinant is zero.

Example 20.11

$$\Delta = \begin{vmatrix} -3 & 5 & 1 \\ 2 & 4 & 7 \\ -2 & 4 & 7 \end{vmatrix}$$

Subtract the bottom row from the middle row:

$$\Delta = \begin{vmatrix} -4 & 5 & 1 \\ 2-2 & 4-4 & 7-7 \\ 2 & 4 & 7 \end{vmatrix} = \begin{vmatrix} -3 & 5 & 1 \\ 0 & 0 & 0 \\ 2 & 4 & 7 \end{vmatrix}$$

Now it can be seen that if we expand about the middle row the value of the determinant is zero.

Considering these last two examples we may deduce the rule:

> If two columns (or two rows) are identical then the value of the determinant is zero.

EXERCISE

20·6

Evaluate the following determinants:

1 $\begin{vmatrix} 2 & 4 & 1 \\ 4 & 5 & 3 \\ 8 & 7 & 2 \end{vmatrix}$

2 $\begin{vmatrix} 2 & 4 & 3 \\ 9 & 6 & 2 \\ 2 & 7 & 5 \end{vmatrix}$

3 $\begin{vmatrix} 8 & 4 & 2 \\ 9 & 6 & 7 \\ 1 & 3 & 5 \end{vmatrix}$

4 $\begin{vmatrix} 1 & 6 & 3 \\ 9 & 8 & 7 \\ 5 & 4 & 2 \end{vmatrix}$

5 $\begin{vmatrix} 5 & 6 & 2 \\ 7 & 1 & 3 \\ 2 & 3 & 4 \end{vmatrix}$

6 $\begin{vmatrix} 4 & 7 & 1 \\ 6 & 3 & 2 \\ 9 & 1 & 8 \end{vmatrix}$

7 $\begin{vmatrix} 1 & 3 & 2 \\ 2 & 1 & 3 \\ -5 & 2 & -1 \end{vmatrix}$

8 $\begin{vmatrix} 4 & 1 & 2 \\ -3 & 1 & -5 \\ 8 & -6 & 11 \end{vmatrix}$

9 $\begin{vmatrix} 3 & 2 & 1 \\ 1 & -1 & -3 \\ 5 & -11 & -3 \end{vmatrix}$

10 $\begin{vmatrix} 3 & 5 & -4 \\ -1 & -3 & 2 \\ 6 & 2 & 8 \end{vmatrix}$

11 $\begin{vmatrix} 4 & 5 & -2 \\ -3 & -2 & -1 \\ 7 & 3 & -3 \end{vmatrix}$

12 $\begin{vmatrix} 3 & 1 & 2 \\ -1 & 4 & 1 \\ -10 & 1 & -5 \end{vmatrix}$

13 $\begin{vmatrix} 2 & 2 & 4 \\ 2 & 4 & 2 \\ 4 & 2 & 2 \end{vmatrix}$

14 $\begin{vmatrix} 1 & 2 & 1 \\ 2 & 1 & 1 \\ 1 & 1 & 2 \end{vmatrix}$

15 $\begin{vmatrix} 2 & 3 & 1 \\ 3 & 1 & 2 \\ 1 & 2 & 3 \end{vmatrix}$

16 $\begin{vmatrix} 3 & 5 & 2 \\ -1 & 7 & -2 \\ 3 & 5 & 2 \end{vmatrix}$

17 $\begin{vmatrix} 3 & 2 & 3 \\ -4 & 1 & -4 \\ 7 & 4 & 7 \end{vmatrix}$

18 $\begin{vmatrix} 3 & 9 & 6 \\ 1 & 3 & 2 \\ 4 & -4 & 1 \end{vmatrix}$

19 $\begin{vmatrix} 2 & 1 & 3 \\ 4 & 5 & 2 \\ 1 & 7 & 0 \end{vmatrix}$

20 $\begin{vmatrix} 37 & 36 & 39 \\ 42 & 29 & 37 \\ -6 & 7 & 3 \end{vmatrix}$

21 $\begin{vmatrix} 3 & 2 & 9 \\ 3 & 3 & 10 \\ 1 & -1 & 10 \end{vmatrix}$

22 $\begin{vmatrix} 10 & 12 & 3 \\ 14 & 18 & 5 \\ 6 & 7 & 2 \end{vmatrix}$

23 $\begin{vmatrix} 5 & 3 & 2 \\ 7 & -2 & 3 \\ 5 & 3 & -4 \end{vmatrix}$

24 $\begin{vmatrix} a & a & (a-1) \\ a & (a+1) & a \\ (1-a) & 1 & a \end{vmatrix}$

Vectors

OBJECTIVES

1 Define a vector.

2 Define the addition of two or more vectors.

3 Resolve a vector into two component parts at right angles to each other.

4 Calculate the magnitude and direction of the resultant of two component parts at right angles to each other.

5 Resolve a system of vectors into component parts and calculate the totals of the horizontal and vertical components.

6 Calculate the magnitude and direction of the resultant of the two total component parts calculated.

7 State the parallelogram law for the addition of any two vectors.

8 Define the result of multiplying a vector by a scalar.

9 Define the inverse vector $(-\mathbf{a})$.

10 Deduce the resultant of vector subtraction $(\mathbf{a} - \mathbf{b})$ and relate the result to the second diagonal of the parallelogram.

11 Solve simple problems involving the addition and subtraction of vectors.

12 Define the unit vectors $\mathbf{i}, \mathbf{j}, \mathbf{k}$ in three dimensions.

13 Express any vector $\mathbf{r}$ in the form $x\mathbf{i} + y\mathbf{j} + z\mathbf{k}$.

14 Define the scalar product of two vectors as $\mathbf{a} \cdot \mathbf{b} = ab \cos \theta$.

15 Deduce that $\mathbf{a} \cdot \mathbf{b} = \mathbf{b} \cdot \mathbf{a}$.

16 Deduce the scalar product in terms of the components of $\mathbf{a}$ and $\mathbf{b}$.

17 Obtain the angle between two vectors given in component form in three dimensions.

18 Derive the expression for work done as $\mathbf{F} \cdot \mathbf{d}$.

19 Define the vector product $\mathbf{a} \times \mathbf{b}$.

20 Deduce that $\mathbf{a} \times \mathbf{b} = -\mathbf{b} \times \mathbf{a}$.

21 Deduce the vector product in terms of the components of $\mathbf{a}$ and $\mathbf{b}$.

22 Express the result of **21** as a determinant.

23 Solve a simple mechanical problem using the vector product, e.g. rigid body rotation.

Vectors

A vector quantity needs both direction and magnitude to describe it fully. A vector may be represented by a section of a straight line. The length of the line represents the magnitude of the vector and the direction of the line represents the direction of the vector (Fig. 21.1).

Fig. **21.1**

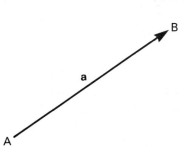

The usual way of naming a vector is to name its end points. The vector in Fig. 21.1 starts at A and ends at B and we write $\overrightarrow{AB}$ which means 'the vector from A to B'.

Sometimes vectors are printed in heavy type; thus the vector in the diagram may be written **AB** or **a**.

Some examples of vector quantities are:

(1) A displacement in a given direction, e.g. 15 metres due west.
(2) A velocity in a given direction, e.g. 40 km/h due north.
(3) A force of 20 newtons acting vertically downwards.

Representing Vectors

It follows that vectors can be represented by an accurately scaled line.

Example 21.1

A man walked a distance of 8 km due east. Draw the vector.

We first choose a suitable scale to represent the magnitude of the vector. In Fig. 21.2 a scale of 1 cm = 2 km has been chosen. We then draw a horizontal line 4 cm long. An arrow is placed on the line to show that the man walked in an easterly direction (and not in a westerly direction). The way in which the arrowhead points gives the sense of the vector.

Fig. **21.2**

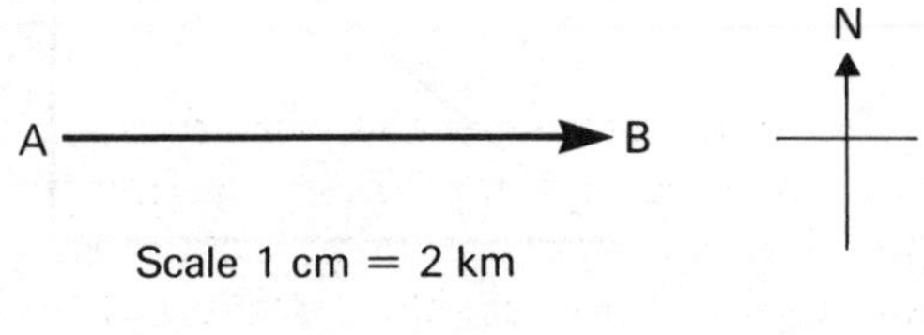

A ➝ B   N ↑

Scale 1 cm = 2 km

Scalars

A scalar quantity is one that is fully defined by magnitude alone. Some examples of scalar quantities are time (e.g. 35 seconds), temperature (8 degrees Celsius) and mass (8 kilograms).

Cartesian Components

Fig. 21.3 shows a vector $\overrightarrow{AB}$ of magnitude 6 units making an angle of 30° to the horizontal. Instead of stating the magnitude of the vector and the angle giving its direction the vector can be defined by splitting it up into its horizontal and vertical components. Thus in the diagram the vector $\overrightarrow{AB}$ can be defined by stating its horizontal component $\overrightarrow{AC}$ and its vertical component $\overrightarrow{CB}$. Vectors $\overrightarrow{AC}$ and $\overrightarrow{CB}$ are said to be the Cartesian components of $\overrightarrow{AB}$.

Fig. **21.3**

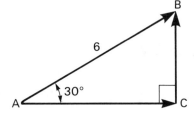

Example 21.2

The vector shown in Fig. 21.4 has a vertical component of 3 units and a horizontal component of 4 units. Calculate its magnitude and direction.

Fig. **21.4**

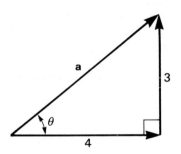

The magnitude of a vector is sometimes called its **modulus**. Thus if the vector is **a** then its modulus is written as $|\mathbf{a}|$ or a. (The quantity $|\mathbf{a}|$ or a is a scalar.)

Using Pythagoras' theorem, the magnitude of $\overrightarrow{AB}$ is

$$|\mathbf{a}| = \sqrt{3^2 + 4^2} = 5 \text{ units}$$

To state its direction we find the size of the angle θ:

$$\tan \theta = \frac{3}{4}$$

 $$\theta = 36.9°$$

Finding the Cartesian Components of a Vector

These may be found by making a scaled drawing or by using trigonometry. Thus in Fig. 21.5

$$\text{Vertical component} = \mathbf{a} \sin \theta$$

$$\text{Horizontal component} = \mathbf{a} \cos \theta$$

Fig. **21.5**

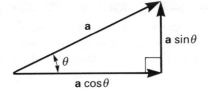

Example 21.3

Find the magnitude of the vertical and horizontal components of the vector shown in Fig. 21.6.

Fig. **21.6**

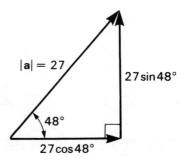

Magnitude of horizontal component $= |\mathbf{a} \cos 48°| = 27 \cos 48° = 18.1$

Magnitude of vertical component $= |\mathbf{a} \sin 48°| = 27 \sin 48° = 20.1$

When a vector is stated in component form it can be written as a column matrix. Thus the vector **a** shown in Fig. 21.7 is $\mathbf{a} = \begin{pmatrix} 5 \\ 4 \end{pmatrix}$.

Fig. **21.7**

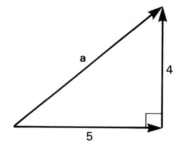

Unit Vectors

Any vector whose magnitude is 1 unit is called a unit vector.

A frame of reference consisting of an origin O and a pair of perpendicular vectors is very similar to the Cartesian xy plane. In Fig. 21.8, **i** is a unit vector in the positive direction of the x-axis and **j** is a unit vector in the positive direction of the y-axis and it is now possible to define a vector in terms of **i** and **j**.

Fig. **21.8**   Fig. **21.9**

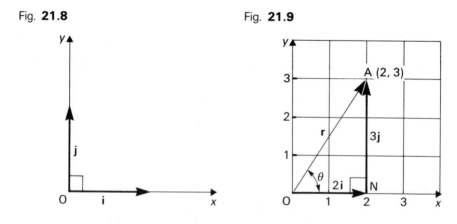

In Fig. 21.9, the point A has the coordinates (2, 3).

Thus $\overrightarrow{ON} = 2\mathbf{i}$ and $\overrightarrow{NA} = 3\mathbf{j}$

The position vector of A(2, 3) is $\mathbf{r} = 2\mathbf{i} + 3\mathbf{j}$.

Example 21.4

Calculate the magnitude and direction of the vector $\mathbf{v} = 3\mathbf{i} + 5\mathbf{j}$.

P is the point with coordinates (3, 5) as shown in Fig. 21.10. The problem is to find the length of OP and the size of the angle θ.

Fig. **21.10**

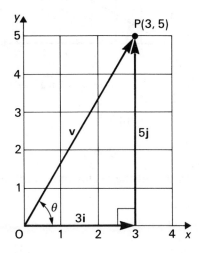

The magnitude of $\mathbf{v}$ is given by the length of OP.

So
$$|\mathbf{v}| = \sqrt{3^2 + 5^2} = 5.83$$

$$\tan \theta = \frac{5}{3}$$

$$\theta = 59.0°$$

So the vector has a magnitude of 5.83 units and it makes an angle of 59.0° with the horizontal.

Example 21.5

Find the magnitude and direction of the vector $\mathbf{a} = \begin{pmatrix} 7 \\ 4 \end{pmatrix}$.

In Fig. 21.11, P is the point (7, 4). The magnitude of $\mathbf{a}$ is given by the length of OP.

Fig. **21.11**

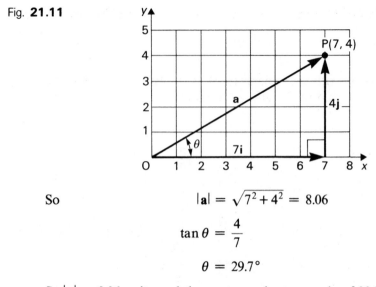

So $\qquad |\mathbf{a}| = \sqrt{7^2 + 4^2} = 8.06$

$$\tan\theta = \frac{4}{7}$$

$$\theta = 29.7°$$

So $|\mathbf{a}| = 8.06$ units and the vector makes an angle of 29.7° to the horizontal.

Example 21.6

Calculate the magnitude and direction of the vector $\mathbf{a} = 4\mathbf{i} - 2\mathbf{j}$.

In Fig. 21.12, P is the point $(4, -2)$. The magnitude of $\mathbf{a}$ is given by

$$|\mathbf{a}| = \sqrt{4^2 + (-2)^2} = 4.47$$

$$\tan\theta = \frac{-2}{4}$$

$$\theta = -26.6°$$

Fig. **21.12**

You should always sketch the vector; otherwise a mistake may be made in the direction of the vector, particularly if a scientific calculator is being used.

EXERCISE

1 Which of the following quantities are vectors?
  (a) quantity of heat
  (b) acceleration
  (c) atomic weight
  (d) centripetal force
  (e) magnetic intensity
  (f) momentum

2 Draw the following vectors, using a convenient scale for each. Mark each of the angles clearly.
  (a) 6 newtons acting at 45° to the horizontal
  (b) 7 m/s acting at 75° to the horizontal
  (c) 20 km acting at −20° to the horizontal
  (d) 4 m/s^2 acting at −30° to the horizontal

3 For each of the vectors shown in Fig. 21.13, find the horizontal and vertical components. Then write each of them in the equivalent component form (i.e. as column matrices).

Fig. **21.13**

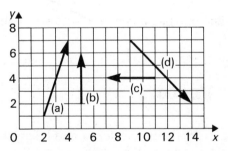

4 (a) Using squared paper draw lines to represent each of the following vectors:

(i) $\begin{pmatrix} 3 \\ 3 \end{pmatrix}$ (ii) $\begin{pmatrix} 4 \\ -5 \end{pmatrix}$ (iii) $\begin{pmatrix} 6 \\ -4 \end{pmatrix}$ (iv) $\begin{pmatrix} -4 \\ -5 \end{pmatrix}$

(v) $\begin{pmatrix} -5 \\ 0 \end{pmatrix}$ (vi) $\begin{pmatrix} 0 \\ 4 \end{pmatrix}$ (vii) $\begin{pmatrix} 3 \\ 0 \end{pmatrix}$ (viii) $\begin{pmatrix} -2 \\ -6 \end{pmatrix}$

(ix) 4**i** + 3**j**   (x) 5**i** − 3**j**
(xi) −4**i** − 3**j**   (xii) −2**i** + 7**j**

(b) By calculation find the magnitude and direction of each of these vectors.

5 If O is the origin and A is the point with coordinates (4, 6),

    (a) Write down in **i, j** form the equation of the vector $\overrightarrow{OA}$

    (b) Work out the magnitude of $\overrightarrow{OA}$ and find its direction by stating the angle that it makes with the horizontal.

Equal Vectors

Two vectors are equal if they have the same magnitude and the same direction. Thus in Fig. 21.14, the vector $\overrightarrow{AB}$ is equal to the vector $\overrightarrow{CD}$ because the length of AB is the same as the length of CD, and AB is parallel to CD.

Fig. **21.14**

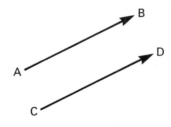

Inverse Vectors

In Fig. 21.15, the two vectors **a** and **b** have the same magnitude but opposite directions. We say that

$$\mathbf{a} = -\mathbf{b}$$

and **b** is said to be the inverse of **a**.

Fig. **21.15**

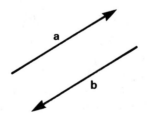

Multiplication of a Vector by a Scalar

If the vector $\overrightarrow{AB}$ is 5 m due north, then the vector $3\overrightarrow{AB}$ is 15 m due north. In general, if n is any number, the vector $n\overrightarrow{AB}$ is a vector in the same direction as $\overrightarrow{AB}$ whose length is n times the length of $\overrightarrow{AB}$.

Addition of Vectors

In Fig. 21.16, A, B and C are three points marked out in a field. A man walks from A to B (i.e. he describes $\overrightarrow{AB}$) and then walks from B to C (i.e. he describes $\overrightarrow{BC}$). Instead the man could have walked from A to C, thus describing the vector $\overrightarrow{AC}$.

Now going from A to C directly has the same result as going from A to C via B. We therefore call $\overrightarrow{AC}$ the **resultant** of the vectors $\overrightarrow{AB}$ and $\overrightarrow{BC}$. We say that the resultant vector $\overrightarrow{AC}$ is the sum of the two vectors $\overrightarrow{AB}$ and $\overrightarrow{BC}$ and we write

$$\overrightarrow{AC} = \overrightarrow{AB} + \overrightarrow{BC}$$

Note carefully that the arrows on the vectors $\overrightarrow{AB}$ and $\overrightarrow{BC}$ follow nose to tail whilst the arrowhead on $\overrightarrow{AC}$ opposes the arrowheads on $\overrightarrow{AB}$ and $\overrightarrow{BC}$. The sense of the resultant $\overrightarrow{AC}$ is often marked with a double arrowhead.

Fig. **21.16**

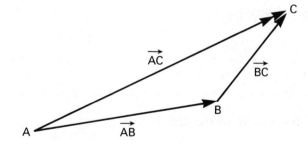

Triangle Law

The triangle law states that if the sides AB and BC of a triangle ABC represent the vectors **c** and **a** respectively, then the third side AC represents the resultant, or the vector sum of **a** and **b**. If AC represents the vector **b** then the vector sum is expressed by the vector equation

$$\mathbf{b} = \mathbf{c} + \mathbf{a}$$

Fig. **21.17**

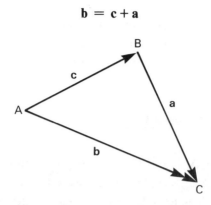

The vector sum of two vectors may be obtained by an accurate scale drawing or by calculation as shown in the following examples.

Example 21.7

Two forces act at a point O as shown in Fig. 21.18. By making an accurate scale drawing, find the resultant force.

Fig. **21.18**   Fig. **21.19**

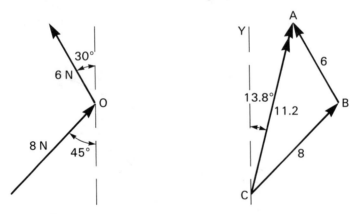

The vector triangle is drawn in Fig. 21.19. The resultant is $\overrightarrow{CA}$ and by measurement it is found to be 11.2 units long. Its direction is given by measuring $\angle ACY$, CY being vertical.

Hence the resultant of the two forces is 11.2 N acting at 13.8° to the vertical.

Example 21.8

Find the resultant of the vectors $\begin{pmatrix} 6 \\ 2 \end{pmatrix}$ and $\begin{pmatrix} 7 \\ 8 \end{pmatrix}$.

The components of the resultant **r** are obtained by adding the two column matrices together.

Thus
$$\mathbf{r} = \begin{pmatrix} 6 \\ 2 \end{pmatrix} + \begin{pmatrix} 7 \\ 8 \end{pmatrix} = \begin{pmatrix} 13 \\ 10 \end{pmatrix}$$

giving
$$|\mathbf{r}| = \sqrt{13^2 + 10^2} = 16.4$$

If θ is the angle which **r** makes with the horizontal, then

$$\tan \theta = \frac{10}{13}$$

$\therefore$
$$\theta = 37.6°$$

Example 21.9

Find the resultant of the vectors $\mathbf{a} = 3\mathbf{i} + 5\mathbf{j}$ and $\mathbf{b} = 4\mathbf{i} + 6\mathbf{j}$.

If **r** is the resultant of **a** and **b** then

$$\mathbf{r} = \mathbf{a} + \mathbf{b}$$
$$= 3\mathbf{i} + 5\mathbf{j} + 4\mathbf{i} + 6\mathbf{j}$$
$$= (3 + 4)\mathbf{i} + (5 + 6)\mathbf{j}$$
$$= 7\mathbf{i} + 11\mathbf{j}$$

giving
$$|\mathbf{r}| = \sqrt{7^2 + 11^2} = 13.0$$

If θ is the angle that **r** makes with the horizontal, then

$$\tan \theta = \frac{11}{7}$$

$\therefore$
$$\theta = 57.5°$$

The Sum of Several Vectors

This is obtained by successive application of the triangle law; a polygon of vectors is formed as shown in Fig. 21.20. Thus

$$r = a + b + c + d$$

Fig. **21.20**

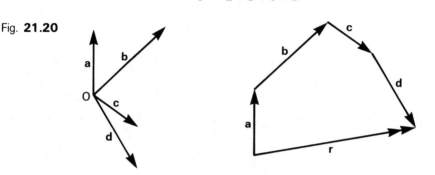

The addition of vectors is commutative, i.e. it does not matter in which order the vectors are added. That is

$$\boxed{a + b = b + a}$$

Also the law for addition is associative, i.e. if **a**, **b** and **c** are three vectors then

$$\boxed{a + (b + c) = (a + b) + c}$$

This means that we could add **a** to the sum of **b** and **c** or we could add the sum of **a** and **b** to **c**, and obtain the same result.

Example 21.10

Calculate the resultant of the three forces shown in Fig. 21.21.

Fig. **21.21**

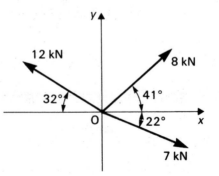

$$\text{Vertical component of } \overrightarrow{OA} = 8 \sin 41° = 5.25 \text{ kN}$$

$$\text{Horizontal component of } \overrightarrow{OA} = 8 \cos 41° = 6.04 \text{ kN}$$

$$\therefore \qquad \overrightarrow{OA} = 6.04\mathbf{i} + 5.25\mathbf{j}$$

$$\text{Vertical component of } \overrightarrow{OB} = -7 \sin 22° = -2.62 \text{ kN}$$

$$\text{Horizontal component of } \overrightarrow{OB} = 7 \cos 22° = 6.49 \text{ kN}$$

$$\therefore \qquad \overrightarrow{OB} = 6.49\mathbf{i} - 2.62\mathbf{j}$$

$$\text{Vertical component of } \overrightarrow{OC} = 12 \sin 32° = 6.36 \text{ kN}$$

$$\text{Horizontal component of } \overrightarrow{OC} = -12 \cos 32° = -10.18 \text{ kN}$$

$$\therefore \qquad \overrightarrow{OC} = -10.18\mathbf{i} + 6.36\mathbf{j}$$

The resultant $\mathbf{r}$ of $\overrightarrow{OA}$, $\overrightarrow{OB}$ and $\overrightarrow{OC}$ is

$$\mathbf{r} = (6.04\mathbf{i} + 5.25\mathbf{j}) + (6.49\mathbf{i} - 2.62\mathbf{j}) + (-10.18\mathbf{i} + 6.36\mathbf{j})$$

$$= 2.35\mathbf{i} + 8.99\mathbf{j}$$

$$|\mathbf{r}| = \sqrt{2.35^2 + 8.99^2} = 9.29$$

$$\tan \theta = \frac{8.99}{2.35}$$

$$\therefore \qquad \theta = 75.4°$$

Hence the resultant force is 9.29 kN making an angle of 75.4° to the horizontal.

Vector Addition Using the Parallelogram Law

Vector addition also follows the parallelogram law which applies to two vectors acting at a point as shown in Fig. 21.22. The vector sum (or resultant) of the two vectors **a** and **b** is obtained by drawing lines parallel to the vectors **a** and **b** to form a parallelogram. The vector sum is then represented by the diagonal **r**. Hence

$$\mathbf{r} = \mathbf{a} + \mathbf{b}$$

Fig. **21.22**

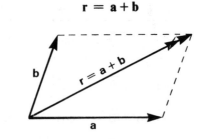

Example 21.11

Find the resultant of the two forces shown in Fig. 21.23.

Fig. **21.23**   Fig. **21.24**

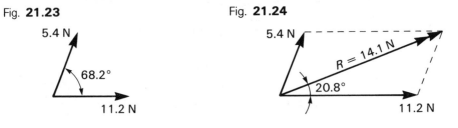

The parallelogram of forces is shown in Fig. 21.24. By measurement the resultant is found to be 14.1 N making an angle of 20.8° to the horizontal.

Subtraction of Vectors

To subtract a vector we add its inverse. Hence

$$\overrightarrow{AC} - \overrightarrow{BC} = \overrightarrow{AC} + \overrightarrow{CB} = \overrightarrow{AB}$$

Example 21.12

If $\mathbf{a} = \begin{pmatrix} 1 \\ 4 \end{pmatrix}$ and $\mathbf{b} = \begin{pmatrix} 5 \\ -1 \end{pmatrix}$ find $\mathbf{a} - \mathbf{b}$.

The two vectors are shown in Fig. 21.25.

Fig. **21.25**

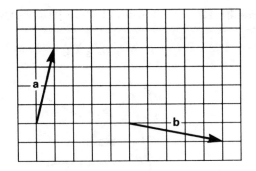

To obtain **−b** we reverse the direction of **b** and we then perform the addition

$$\mathbf{r} = \mathbf{a} + (-\mathbf{b})$$

Fig. **21.26**

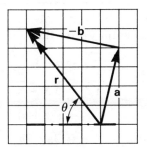

The graphical solution is given in Fig. 21.26 where it will be seen that

$$\mathbf{r} = \mathbf{a} - \mathbf{b} = 6.4 \text{ units with } \theta = 51.3°$$

The same result may be obtained by subtracting the column matrices as follows:

$$\mathbf{r} = \mathbf{a} - \mathbf{b} = \begin{pmatrix} 1 \\ 4 \end{pmatrix} - \begin{pmatrix} 5 \\ -1 \end{pmatrix} = \begin{pmatrix} -4 \\ 5 \end{pmatrix}$$

$$|\mathbf{r}| = \sqrt{(-4)^2 + 5^2} = 6.4$$

$$\tan \theta = \frac{5}{-4}$$

∴ $$\theta = -51.3°$$

If preferred the vectors can be expressed in the **i, j** form as follows:

$$\mathbf{a} = \mathbf{i} + 4\mathbf{j} \text{ and } \mathbf{b} = 5\mathbf{i} - \mathbf{j}$$

$$\mathbf{r} = \mathbf{a} - \mathbf{b} = \mathbf{i} + 4\mathbf{j} - (5\mathbf{i} - \mathbf{j})$$

$$= -4\mathbf{i} + 5\mathbf{j}$$

which gives the same results as before.

Example 21.13

(a) In Fig. 21.27, $\overrightarrow{PX} = \mathbf{a}$, $\overrightarrow{XQ} = \mathbf{b}$, $\overrightarrow{YP} = 2\mathbf{a}$ and $\overrightarrow{QZ} = 2\mathbf{b}$. Find YZ in terms of $\mathbf{a}$ and $\mathbf{b}$ and show that PQ and YZ are parallel.

Fig. **21.27**

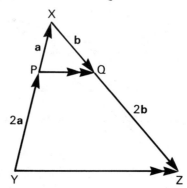

Looking at the diagram we see that

$$\overrightarrow{PQ} = \mathbf{a} + \mathbf{b}$$

and

$$\overrightarrow{YZ} = \overrightarrow{YX} + \overrightarrow{XZ}$$

$$= (2\mathbf{a} + \mathbf{a}) + (\mathbf{b} + 2\mathbf{b})$$

$$= 3\mathbf{a} + 3\mathbf{b}$$

$$= 3(\mathbf{a} + \mathbf{b})$$

Hence $\overrightarrow{YZ}$ has a magnitude which is three times greater than that of $\overrightarrow{PQ}$. Also PQ and YZ must be parallel since $\overrightarrow{PQ}$ and $\overrightarrow{YZ}$ have the same direction.

Example 21.14

Fig. **21.28**

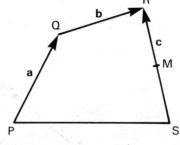

In Fig. 21.28, M is the mid-point of RS, $\overrightarrow{PQ} = \mathbf{a}$, $\overrightarrow{QR} = \mathbf{b}$ and $\overrightarrow{MR} = \mathbf{c}$.

Express each of the following in terms of $\mathbf{a}$, $\mathbf{b}$ and $\mathbf{c}$:

(a) $\overrightarrow{SR}$ (b) $\overrightarrow{PR}$ (c) $\overrightarrow{QM}$ (d) $\overrightarrow{PM}$

(a) $\overrightarrow{SR} = 2\overrightarrow{MR} = 2c$

(b) $\overrightarrow{PR} = \overrightarrow{PQ} + \overrightarrow{QR} = a + b$ (Fig. 21.29)

Fig. **21.29**

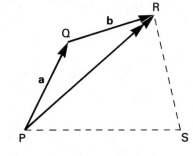

(c) In $\triangle$QRM (Fig. 21.30), $\overrightarrow{QR}$ is the resultant of $\overrightarrow{QM}$ and $\overrightarrow{MR}$.

$\therefore \quad \overrightarrow{QR} = \overrightarrow{QM} + \overrightarrow{MR}$

Thus $\overrightarrow{QM} = \overrightarrow{QR} - \overrightarrow{MR} = b - c$

Fig. **21.30**

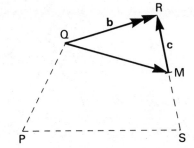

(d) In $\triangle$PRM (Fig. 21.31) $\overrightarrow{PR}$ is the resultant of $\overrightarrow{PM}$ and $\overrightarrow{MR}$.

$\therefore \quad \overrightarrow{PR} = \overrightarrow{PM} + \overrightarrow{MR}$

Thus $\overrightarrow{PM} = \overrightarrow{PR} - \overrightarrow{MR} = a + b - c$

Fig. **21.31**

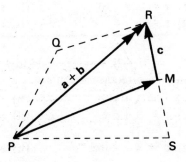

EXERCISE

21·2

1 For the pairs of vectors shown in Fig. 21.32 find their resultant:
   (a) by making a scale drawing
   (b) by calculation

Fig. **21.32**

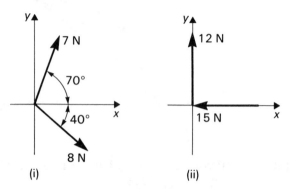

(i)

(ii)

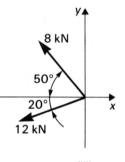

(iii)

2 If $\mathbf{a} = \begin{pmatrix} 3 \\ 5 \end{pmatrix}$, $\mathbf{b} = \begin{pmatrix} 7 \\ 3 \end{pmatrix}$ and $\mathbf{c} = \begin{pmatrix} -3 \\ -2 \end{pmatrix}$, calculate $|\mathbf{r}|$ and its direction when:

   (a) $\mathbf{r} = \mathbf{a} + \mathbf{b}$ (b) $\mathbf{r} = \mathbf{a} - \mathbf{b}$
   (c) $\mathbf{r} = 2\mathbf{a}$ (d) $\mathbf{r} = 2\mathbf{a} - \mathbf{b}$
   (e) $\mathbf{r} = 2(\mathbf{a} + \mathbf{b})$ (f) $\mathbf{r} = \mathbf{a} - \mathbf{c}$
   (g) $\mathbf{r} = \mathbf{a} + \mathbf{b} - \mathbf{c}$ (h) $\mathbf{r} = \mathbf{a} - 2\mathbf{b} + \mathbf{c}$

3 If $\mathbf{x} = 3\mathbf{i} + 2\mathbf{j}$, $\mathbf{y} = 5\mathbf{i} - 3\mathbf{j}$ and $\mathbf{z} = 4\mathbf{i} - 2\mathbf{j}$, calculate $|\mathbf{r}|$ and state its direction when:

   (a) $\mathbf{r} = \mathbf{x} + \mathbf{y}$ (b) $\mathbf{r} = \mathbf{x} - \mathbf{y}$
   (c) $\mathbf{r} = \mathbf{x} + \mathbf{z}$ (d) $\mathbf{r} = \mathbf{x} - \mathbf{z}$
   (e) $\mathbf{r} = 2\mathbf{y}$ (f) $\mathbf{r} = 2\mathbf{x} - 3\mathbf{y}$
   (g) $\mathbf{r} = \mathbf{x} + \mathbf{y} + \mathbf{z}$ (h) $\mathbf{r} = 2\mathbf{x} + 3\mathbf{y} - 4\mathbf{z}$

4 Fig. 21.33 shows the vectors **a** and **b**. Find, by making scale drawings, the magnitude and direction of:

(a) **r** = **a** + **b**   (b) **r** = **a** − **b**

Fig. **21.33**

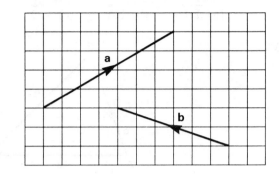

5 In Fig. 21.34 ABCD is a parallelogram. $\overrightarrow{AB}$ = **a** and $\overrightarrow{BC}$ = **b**. Express in terms of **a** and **b**:

(a) $\overrightarrow{AD}$ (b) $\overrightarrow{CD}$ (c) $\overrightarrow{AC}$

Fig. **21.34**

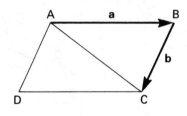

6 In Fig. 21.35, XYZ is an isosceles triangle. XY and YZ are the equal sides. YW is a perpendicular from Y to XZ. If $\overrightarrow{XY}$ = **a** and $\overrightarrow{YZ}$ = **b**, express in terms of **a** and **b**:

(a) $\overrightarrow{XZ}$ (b) $\overrightarrow{ZW}$ (c) $\overrightarrow{YW}$

Fig. **21.35**

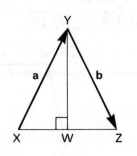

7 In Fig. 21.36, ABC is a triangle in which $\overrightarrow{AC} = \mathbf{a}$, $\overrightarrow{BC} = \mathbf{b}$ and $\overrightarrow{AD} = \mathbf{c}$, D being the mid-point of AB. Express in terms of $\mathbf{a}$, $\mathbf{b}$ and $\mathbf{c}$:

(a) $\overrightarrow{DB}$ (b) $\overrightarrow{BA}$ (c) $\overrightarrow{CD}$

Fig. **21.36**

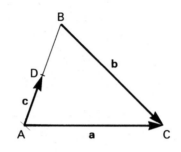

8 In Fig. 21.37, D and E are the mid-points of AB and AC respectively. $\overrightarrow{AD} = \mathbf{a}$ and $\overrightarrow{AE} = \mathbf{b}$.

(a) Express in terms of $\mathbf{a}$ and $\mathbf{b}$:
   (i) $\overrightarrow{ED}$ (ii) $\overrightarrow{DE}$ (iii) $\overrightarrow{BD}$ (iv) $\overrightarrow{AB}$
   (v) $\overrightarrow{CE}$ (vi) $\overrightarrow{AC}$
(b) Show that DE and BC are parallel.

Fig. **21.37**

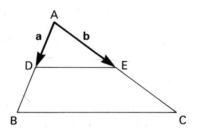

9 In Fig. 21.38, ABCD is a trapezium with AB parallel to DC and DC = 2AB. If $\overrightarrow{DA} = \mathbf{a}$ and $\overrightarrow{AB} = \mathbf{b}$, express in terms of $\mathbf{a}$ and $\mathbf{b}$:

(a) $\overrightarrow{DB}$ (b) $\overrightarrow{DC}$ (c) $\overrightarrow{BC}$

Fig. **21.38**

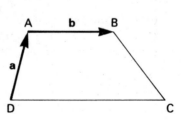

10 In Fig. 21.39 ABCDEF is a regular hexagon. If $\overrightarrow{AF}$ = **a**, $\overrightarrow{AB}$ = **b** and $\overrightarrow{BC}$ = **c**, express each of the following in terms of **a**, **b** and **c**:

(a) $\overrightarrow{DC}$ (b) $\overrightarrow{DE}$ (c) $\overrightarrow{FE}$ (d) $\overrightarrow{FC}$ (e) $\overrightarrow{AE}$ (f) $\overrightarrow{AD}$

Fig. **21.39**

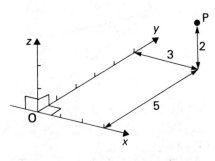

Vectors in Three Dimensions

Fig. 21.40 shows a Cartesian frame of reference which consists of a fixed point O, the origin, and three mutually perpendicular axes Ox, Oy and Oz. A point is located within the frame by stating its directed distances along the positive x-, y- and z-axes respectively.

Fig. **21.40**

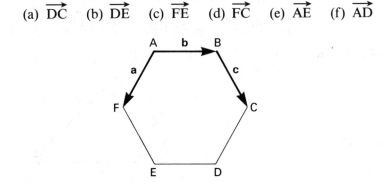

In Fig. 21.40 the point P whose coordinates are (3, 5, 2) is

3 units from O in the x-direction
5 units from O in the y-direction
2 units from O in the z-direction

Now if **i** is a unit vector in the x-direction
and **j** is a unit vector in the y-direction
and **k** is a unit vector in the z-direction

then the position vector $\overrightarrow{OP}$ for the point (3, 5, 2) is given by

$$\overrightarrow{OP} = 3\mathbf{i} + 5\mathbf{j} + 2\mathbf{k}$$

The vector $\overrightarrow{OP}$ may also be written as the column vector $\begin{pmatrix} 3 \\ 5 \\ 2 \end{pmatrix}$.

The magnitude of the vector

$$|\overrightarrow{OP}| = \sqrt{3^2 + 5^2 + 2^2} \quad \text{using Pythagoras' theorem twice.}$$

$$= 6.16$$

Example 21.15

A triangle ABC has its vertices at the points A(4, −2, 8), B(6, −4, 10) and C(−2, 12, 4). Find in the form $a\mathbf{i} + b\mathbf{j} + c\mathbf{k}$, the vectors $\overrightarrow{AB}$, $\overrightarrow{BC}$ and $\overrightarrow{CA}$ and hence find the lengths of the sides of the triangle.

When two or more points are to be shown in a diagram it is better not to show the axes because these will reduce the clarity of the diagram. However, as shown in Fig. 21.41, the origin should be marked since this gives a reference point.

Fig. **21.41**

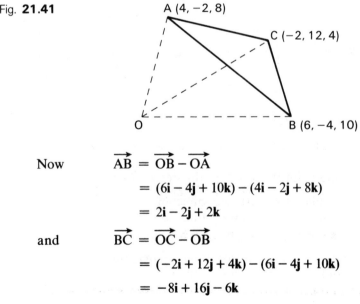

Now $\qquad \overrightarrow{AB} = \overrightarrow{OB} - \overrightarrow{OA}$

$\qquad\qquad = (6\mathbf{i} - 4\mathbf{j} + 10\mathbf{k}) - (4\mathbf{i} - 2\mathbf{j} + 8\mathbf{k})$

$\qquad\qquad = 2\mathbf{i} - 2\mathbf{j} + 2\mathbf{k}$

and $\qquad \overrightarrow{BC} = \overrightarrow{OC} - \overrightarrow{OB}$

$\qquad\qquad = (-2\mathbf{i} + 12\mathbf{j} + 4\mathbf{k}) - (6\mathbf{i} - 4\mathbf{j} + 10\mathbf{k})$

$\qquad\qquad = -8\mathbf{i} + 16\mathbf{j} - 6\mathbf{k}$

and $\qquad \overrightarrow{CA} = \overrightarrow{OA} - \overrightarrow{OC}$

$$= (4i - 2j + 8k) - (-2i + 12j + 4k)$$

$$= 6i - 14j + 4k$$

Thus $\qquad AB = |\overrightarrow{AB}| = \sqrt{2^2 + (-2)^2 + 2^2} = 3.46$

and $\qquad BC = |\overrightarrow{BC}| = \sqrt{(-8)^2 + 16^2 + (-6)^2} = 18.87$

and $\qquad CA = |\overrightarrow{CA}| = \sqrt{6^2 + (-14)^2 + 4^2} = 15.75$

EXERCISE

21·3

1 Write down in the form $ai + bj + ck$, the vector represented by $\overrightarrow{OP}$ if the point P has the coordinates:

(a) $(2, 3, 7)$ (b) $1, -3, 4)$ (c) $0, 1, -3)$

2 $\overrightarrow{OP}$ represents the vector **r**. Find the coordinates of P when:

(a) $r = 4i - 5j + 3k$ (b) $r = i + 3j$ (c) $r = j - k$

3 Find the length of the line OP when P is the point:

(a) $(3, -1, 5)$ (b) $(3, 0, 6)$ (c) $(-2, -2, 3)$

4 Calculate the modulus of the vector **v** when

(a) $v = 3i - 4j + 5k$ (b) $v = 7i + 3j - 4k$

5 If $a = 2i + j + k$, $b = i - j + 4k$ and $c = -2i + 3j + 5k$, write down:

(a) $a + b$ (b) $b - c$ (c) $a + b + c$ (d) $a - b - c$
(e) $3a + 2b - 4c$

6 The triangle ABC has vertices at the points $A(-3, 9, 0)$, $B(-9, 0, 21)$ and $C(-3, 6, 9)$.

(a) Write down in the form $ai + bj + ck$ the vectors representing $\overrightarrow{AB}$, $\overrightarrow{AC}$ and $\overrightarrow{CB}$.

(b) Calculate the lengths of the sides of the triangle.

7 If $a = \begin{pmatrix} 3 \\ 2 \\ 1 \end{pmatrix}$, $b = \begin{pmatrix} 1 \\ 1 \\ 1 \end{pmatrix}$ and $c = \begin{pmatrix} 0 \\ 0 \\ 2 \end{pmatrix}$ find:

(a) $a + b - c$ (b) $2a - b + 3c$

Scalar Product

Consider a trolley running on rails as shown in Fig. 21.42. It moves in the x-direction under the application of the force **F** which acts at an angle θ to the direction of travel. We want to find the work done by the force when the trolley moves a distance s in the x-direction.

Fig. **21.42**

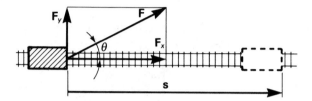

To do this we resolve **F** into its Cartesian components $\mathbf{F}_x$ and $\mathbf{F}_y$. The work done is, by definition, the product of the force in the direction of motion and the distance travelled. Hence the work done by **F** is

$$W = F_x s$$

and since

$$F_x = F \cos \theta$$

$$W = Fs \cos \theta$$

Now work done is a scalar quantity and the product obtained when force is multiplied by distance is called a **scalar product** or a **dot product** because it is usually written:

$$W = \mathbf{F}_x \cdot \mathbf{s} = Fs \cos \theta$$

Note that the work done by $\mathbf{F}_y$ is zero because there is no motion in a vertical direction.

In general:

> The scalar product of two vectors is the product of their magnitudes and the cosine of the angle between their directions.

Thus if **a** and **b** are two vectors and the angle between their directions is θ (Fig. 21.43), then their scalar product is defined as

$$\mathbf{a} \cdot \mathbf{b} = |\mathbf{a}||\mathbf{b}| \cos \theta = ab \cos \theta$$

Fig. **21.43**

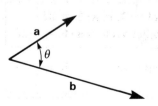

Example 21.16

In Fig. 21.44, F represents the force of gravity acting on a body which slides down an inclined plane. If $F = 5\,\text{N}$ and the plane is inclined at 30° to the horizontal, find the mechanical work done by F when it slides 20 m down the plane.

$$W = \mathbf{F} \cdot \mathbf{s} = Fs \cos \theta = 5 \times 20 \times \cos 60° = 50\,\text{N}\,\text{m}$$

Fig. **21.44**

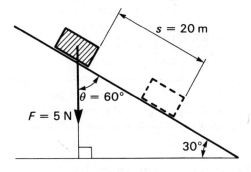

$s = 20\,\text{m}$

$\theta = 60°$

$F = 5\,\text{N}$

30°

Properties of the Scalar Product

Parallel Vectors

If the vectors **a** and **b** are parallel,
then in Fig. 21.45 $\mathbf{a} \cdot \mathbf{b} = ab \cos 0 = ab$ (since $\cos 0 = 1$)
and in Fig. 21.46 $\mathbf{a} \cdot \mathbf{b} = ab \cos \pi = -ab$ (since $\cos \pi = -1$)

Fig. **21.45**   Fig. **21.46**

                  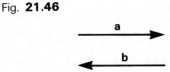

That is:

> for like parallel vectors $\mathbf{a} \cdot \mathbf{b} = ab$
> for unlike parallel vectors $\mathbf{a} \cdot \mathbf{b} = -ab$

In the special case when $\mathbf{a} = \mathbf{b}$ then

$$\mathbf{a} \cdot \mathbf{b} = \mathbf{a} \cdot \mathbf{a} = a^2$$

Perpendicular Vectors

If the vectors $\mathbf{a}$ and $\mathbf{b}$ are perpendicular then $\mathbf{a} \cdot \mathbf{b} = ab \cos \dfrac{\pi}{2}$ and since $\cos \dfrac{\pi}{2} = 0$,

$$\mathbf{a} \cdot \mathbf{b} = 0 \quad \text{for vectors at right angles to each other}$$

Cartesian Unit Vectors

Applying the above results we have:

$$\mathbf{i} \cdot \mathbf{i} = \mathbf{j} \cdot \mathbf{j} = \mathbf{k} \cdot \mathbf{k} = 1 \quad \text{for parallel vectors}$$
$$\mathbf{i} \cdot \mathbf{j} = \mathbf{j} \cdot \mathbf{k} = \mathbf{k} \cdot \mathbf{i} = 0 \quad \text{for perpendicular vectors}$$

The Scalar Product is Commutative

Now $\mathbf{a} \cdot \mathbf{b} = ab \cos \theta$ and $\mathbf{b} \cdot \mathbf{a} = ba \cos \theta$

But $ab \cos \theta = ba \cos \theta$

Hence

$$\mathbf{a} \cdot \mathbf{b} = \mathbf{b} \cdot \mathbf{a}$$

The Scalar Product is Distributive Across Addition

$$\mathbf{a} \cdot (\mathbf{b} + \mathbf{c}) = \mathbf{a} \cdot \mathbf{b} + \mathbf{a} \cdot \mathbf{c}$$

Calculation of $a \cdot b$ when Unit Vectors Are Used

Using the properties of the scalar product,

if $\qquad \mathbf{a} = x_1 \mathbf{i} + y_1 \mathbf{j} + z_1 \mathbf{k}$

and $\qquad \mathbf{b} = x_2 \mathbf{i} + y_2 \mathbf{j} + z_2 \mathbf{k}$

then $\mathbf{a} \cdot \mathbf{b} = (x_1\mathbf{i} + y_1\mathbf{j} + z_1\mathbf{k}) \cdot (x_2\mathbf{i} + y_2\mathbf{j} + z_2\mathbf{k})$

$$= (x_1x_2\mathbf{i} \cdot \mathbf{i} + y_1y_2\mathbf{j} \cdot \mathbf{j} + z_1z_2\mathbf{k} \cdot \mathbf{k})$$
$$+ (x_1y_2\mathbf{i} \cdot \mathbf{j} + y_1z_2\mathbf{j} \cdot \mathbf{k} + z_1x_2\mathbf{k} \cdot \mathbf{i})$$
$$+ (y_1x_2\mathbf{j} \cdot \mathbf{i} + z_1y_2\mathbf{k} \cdot \mathbf{j} + x_1z_2\mathbf{i} \cdot \mathbf{k})$$

$$= (x_1x_2 + y_1y_2 + z_1z_2) + 0 + 0$$

by using the results obtained above.

So

$$\boxed{(x_1\mathbf{i} + y_1\mathbf{j} + z_1\mathbf{k}) \cdot (x_2\mathbf{i} + y_2\mathbf{j} + z_2\mathbf{k}) = x_1x_2 + y_1y_2 + z_1z_2}$$

Example 21.17

If $\mathbf{a} = 2\mathbf{i} - 3\mathbf{j} + 4\mathbf{k}$ and $\mathbf{b} = \mathbf{i} + 3\mathbf{j} - 2\mathbf{k}$, find $\mathbf{a} \cdot \mathbf{b}$.

$$\mathbf{a} \cdot \mathbf{b} = (2)(1) + (-3)(3) + (4)(-2) = -15$$

The Angle Between Two Vectors

As shown below, the scalar product may be used to find the angle between two vectors.

Since $\mathbf{a} \cdot \mathbf{b} = |\mathbf{a}||\mathbf{b}| \cos \theta = ab \cos \theta$

$$\boxed{\theta = \arccos \frac{\mathbf{a} \cdot \mathbf{b}}{|\mathbf{a}||\mathbf{b}|} = \arccos \frac{\mathbf{a} \cdot \mathbf{b}}{ab}}$$

Example 21.18

Find the angle between the vectors $\mathbf{a}$ and $\mathbf{b}$ defined in Example 21.17.

From Example 21.17 we have $\mathbf{a} \cdot \mathbf{b} = -15$.

Also $\qquad |\mathbf{a}| = \sqrt{2^2 + (-3)^2 + 4^2} = \sqrt{29} = 5.39$

and $\qquad |\mathbf{b}| = \sqrt{1^2 + 3^2 + (-2)^2} = \sqrt{14} = 3.74$

Thus $\qquad |\mathbf{a}||\mathbf{b}| = 5.39 \times 3.74 = 20.16$

giving $\qquad \theta = \arccos \dfrac{-15}{20.16} = 138.1°$

EXERCISE
21·4

1 The magnitudes and angle between two vectors, **a** and **b**, are given below. Calculate their scalar product:

(a) $a = 3, b = 2, \theta = \dfrac{\pi}{3}$

(b) $a = 4, b = 8, \theta = 0°$

(c) $a = 2, b = 8, \theta = 45°$

(d) $a = 2.75, b = 3.50, \theta = 120°$

2 Calculate $\mathbf{a} \cdot \mathbf{b}$ if $\mathbf{a} = 4\mathbf{i} - 3\mathbf{j} + \mathbf{k}$ and $\mathbf{b} = 2\mathbf{i} + 5\mathbf{j} - 3\mathbf{k}$.

3 In a triangle ABC, $\overrightarrow{AB} = -\mathbf{i} + 4\mathbf{j}$ and $\overrightarrow{BC} = \mathbf{i} + 2\mathbf{j} + 3\mathbf{k}$. Find the size of the angle ABC.

4 Calculate $\mathbf{a} \cdot \mathbf{b}$ when:

(a) $\mathbf{a} = (3, -1, 4)$ and $\mathbf{b} = (-1, 2, 5)$
(b) $\mathbf{a} = (-1, 8, -4)$ and $\mathbf{b} = (6, 3, 5)$
(c) $\mathbf{a} = (2, -3, 2)$ and $\mathbf{b} = (-2, -2, -2)$

5 A force $\mathbf{F} = (0, 8\,\text{N})$ is applied to a body and moves it through a distance **s** metres such that:

(a) $\mathbf{s} = (4\,\text{m}, 4\,\text{m})$ (b) $\mathbf{s} = (3\,\text{m}, 2\,\text{m})$ (c) $\mathbf{s} = (5\,\text{m}, 0)$.
In each case calculate the work done by the force.

6 Which of the following pairs of vectors are perpendicular?

(a) $\mathbf{a} = (0, -2, 2)$, $\mathbf{b} = (2, 0, 0)$
(b) $\mathbf{a} = (2, -4, 1)$, $\mathbf{b} = (-2, 4, 2)$
(c) $\mathbf{a} = (4, -3, 1)$, $\mathbf{b} = (-1, -2, -2)$

7 Calculate the angle between the vectors **a** and **b** when:

(a) $\mathbf{a} = (2, -3, 4)$ and $\mathbf{b} = (3, 2, -1)$
(b) $\mathbf{a} = (-3, -2, -1)$ and $\mathbf{b} = (3, 4, 5)$
(c) $\mathbf{a} = (-6, 6, 3)$ and $\mathbf{b} = (0, 9, 0)$

Torque

Consider the torque (or couple) C resulting from the application of a force F to a body at a point a distance r from the axis of rotation.

Fig. **21.47**

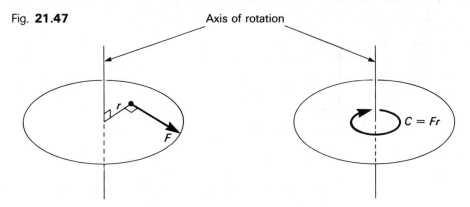

Axis of rotation

$C = Fr$

Fig. 21.47 shows the case when F is at right angles to r. Here $C = Fr$.

Fig. **21.48**

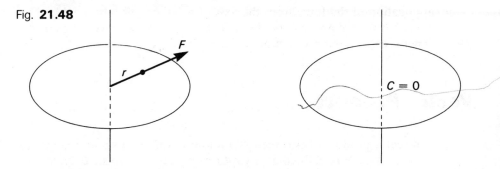

$C = 0$

Fig. 21.48 shows the special case when the force F is in line with r. In this case there is no turning effect and $C = 0$.

Fig. **21.49**

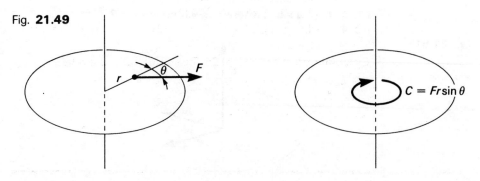

$C = Fr \sin \theta$

Fig. 21.49 shows a more general case where F is inclined to r at an angle θ. Only the component of F at right angles to r, namely $F \sin \theta$, will have any turning effect and so $C = Fr \sin \theta$.

Torque may be represented by a vector in the direction of the axis of rotation. The sense of the vector is determined by using the 'right-handed screw' or 'corkscrew' rule — see Fig. 21.50.

Fig. **21.50**

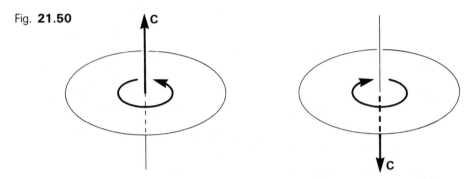

The force may also be represented as a vector **F**. A third vector **r** is known as the position vector which gives the distance of the point of application of the force from the axis of rotation, and also enables the angle between **F** and **r** to be defined. This idea may be extended to the general case of the product of two vectors as shown below.

Vector Product

The vector product of two vectors **a** and **b** is written $\mathbf{a} \times \mathbf{b}$ and is read as 'a cross b'. It is defined as a vector of magnitude $ab \sin \theta$ in a direction perpendicular to the plane containing **a** and **b**, the sense being determined (as explained previously) when turning from **a** to **b** (Fig. 21.51). Here

$$\mathbf{c} = \mathbf{a} \times \mathbf{b} \quad \text{where} \quad |\mathbf{a} \times \mathbf{b}| = ab \sin \theta$$

Fig. **21.51**

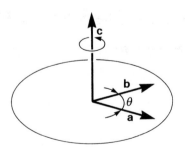

Fig. **21.52**

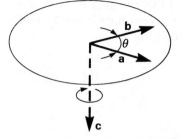

In Fig. 21.52 we have a vector product **b** × **a**, the sense being determined when turning from **b** to **a**. Here **c** = **b** × **a**.

It follows that:

$$\mathbf{a} \times \mathbf{b} = -\mathbf{b} \times \mathbf{a}$$

showing that a vector product is *not* commutative.

However, the distributive laws do apply. Thus:

$$\mathbf{a} \times (\mathbf{b} + \mathbf{c}) = \mathbf{a} \times \mathbf{b} + \mathbf{a} \times \mathbf{c}$$

Also if n is a scalar, then:

$$n\mathbf{a} \times \mathbf{b} = \mathbf{a} \times n\mathbf{b} = n(\mathbf{a} \times \mathbf{b})$$

Example 21.19

Given the two vectors **a** and **b** whose magnitudes are $a = 7$ units and $b = 4$ units with an angle $\theta = 30°$ between them, find the magnitude of **c** = **a** × **b**.

$$|\mathbf{c}| = ab \sin 30° = 7 \times 4 \times 0.5 = 14 \text{ units}$$

Special Cases of the Vector Product

Parallel Vectors

If two vectors are parallel then the angle θ between them is 0 if they are like vectors, and π if they are unlike vectors. In either case $\sin \theta = 0$ and hence

$$|\mathbf{c}| = |\mathbf{a} \times \mathbf{b}| = ab \sin \theta = 0$$

The converse of this statement is also true. If the vector product of two vectors **a** and **b** is zero then it can be concluded that **a** and **b** are parallel. It follows that

$$\mathbf{a} \times \mathbf{b} = 0 \quad \text{for parallel vectors}$$

Perpendicular Vectors

The angle between two perpendicular vectors **a** and **b** is 90° and since $\sin 90° = 1$,

$$|\mathbf{a} \times \mathbf{b}| = ab \sin 90° = ab.$$

Thus:

$$|\mathbf{a} \times \mathbf{b}| = ab \quad \text{for vectors at right angles to each other}$$

Cartesian Unit Vectors

Applying the above results we have (Fig. 21.53):

Fig. **21.53**

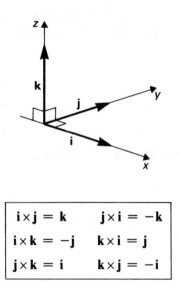

$$
\begin{array}{ll}
\mathbf{i} \times \mathbf{j} = \mathbf{k} & \mathbf{j} \times \mathbf{i} = -\mathbf{k} \\
\mathbf{i} \times \mathbf{k} = -\mathbf{j} & \mathbf{k} \times \mathbf{i} = \mathbf{j} \\
\mathbf{j} \times \mathbf{k} = \mathbf{i} & \mathbf{k} \times \mathbf{j} = -\mathbf{i}
\end{array}
$$

Components of the Vector Product

The vectors **a** and **b** expressed in component form are:

$$\mathbf{a} = a_x\mathbf{i} + a_y\mathbf{j} + a_z\mathbf{k}$$

$$\mathbf{b} = b_x\mathbf{i} + b_y\mathbf{j} + b_z\mathbf{k}$$

Expanding the cross product of **a** and **b** according to the distributive law gives:

$$\mathbf{a} \times \mathbf{b} = (a_x\mathbf{i} + a_y\mathbf{j} + a_z\mathbf{k}) \times (b_x\mathbf{i} + b_y\mathbf{j} + b_z\mathbf{k})$$

$$= \quad a_xb_x(\mathbf{i} \times \mathbf{i}) + a_xb_y(\mathbf{i} \times \mathbf{j}) + a_xb_z(\mathbf{i} \times \mathbf{k})$$

$$+ a_yb_x(\mathbf{j} \times \mathbf{i}) + a_yb_y(\mathbf{j} \times \mathbf{j}) + a_yb_z(\mathbf{j} \times \mathbf{k})$$

$$+ a_zb_x(\mathbf{k} \times \mathbf{i}) + a_zb_y(\mathbf{k} \times \mathbf{j}) + a_zb_z(\mathbf{k} \times \mathbf{k})$$

Using the relationships for the vector products of unit vectors given above we obtain:

$$\boxed{\mathbf{a} \times \mathbf{b} = (a_yb_z - a_zb_y)\mathbf{i} + (a_zb_x - a_xb_z)\mathbf{j} + (a_xb_y - a_yb_x)\mathbf{k}}$$

This vector product may be written in determinant form as follows

$$\mathbf{a} \times \mathbf{b} = \begin{vmatrix} \mathbf{i} & \mathbf{j} & \mathbf{k} \\ a_x & a_y & a_z \\ b_x & b_y & b_z \end{vmatrix}$$

Example 21.20

If $\mathbf{a} = 2\mathbf{i}$ and $\mathbf{b} = 4\mathbf{j}$ find $\mathbf{a} \times \mathbf{b}$.

We are given $a_x = 2$ and $b_y = 4$, all the other coefficients being zero.

Thus
$$\mathbf{a} \times \mathbf{b} = a_xb_y\mathbf{k} = 2 \times 4\mathbf{k} = 8\mathbf{k}$$

Example 21.21

If $\mathbf{a} = \mathbf{i} + \mathbf{j} - \mathbf{k}$ and $\mathbf{b} = 2\mathbf{i} - \mathbf{j} + \mathbf{k}$ find $\mathbf{a} \times \mathbf{b}$.

We are given $a_x = 1, \ a_y = 1, \quad a_z = -1$

and $b_x = 2, \ b_y = -1, \ b_z = 1$

Hence
$$\mathbf{a} \times \mathbf{b} = [1 \times 1 - (-1) \times (-1)]\mathbf{i}$$
$$- [1 \times 1 - (-1) \times 2]\mathbf{j}$$
$$- [1 \times (-1) - 1 \times 2]\mathbf{k}$$
$$= 0\mathbf{i} - 3\mathbf{j} - 3\mathbf{k}$$
$$= -3\mathbf{j} - 3\mathbf{k}$$

Alternatively, using a determinant,

$$\mathbf{a} \times \mathbf{b} = \begin{vmatrix} \mathbf{i} & \mathbf{j} & \mathbf{k} \\ 1 & 1 & -1 \\ 2 & -1 & 1 \end{vmatrix}$$

$$= -3\mathbf{j} - 3\mathbf{k} \text{ (as before)}$$

Example 21.22

Find the angle between the vectors **a** and **b** given that
$\mathbf{a} = (1, -1, 3)$ and $\mathbf{b} = (2, 1, -1)$.

We know that $\qquad |\mathbf{a} \times \mathbf{b}| = ab \sin \theta$

from which $\qquad \sin \theta = \dfrac{|\mathbf{a} \times \mathbf{b}|}{ab}$ $\qquad$ [1]

We are given $\qquad \mathbf{a} \times \mathbf{b} = (\mathbf{i} - \mathbf{j} + 3\mathbf{k}) \times (2\mathbf{i} + \mathbf{j} - \mathbf{k})$

$$= \begin{vmatrix} \mathbf{i} & \mathbf{j} & \mathbf{k} \\ 1 & -1 & 3 \\ 2 & 1 & -1 \end{vmatrix}$$

$$= -2\mathbf{i} + 7\mathbf{j} + 3\mathbf{k}$$

So $\qquad |\mathbf{a} \times \mathbf{b}| = \sqrt{(-2)^2 + (7)^2 + (3)^2}$

$$= 7.87$$

and $\qquad |\mathbf{a}| = a = \sqrt{(2)^2 + (-1)^2 + (3)^2}$

$$= 3.74$$

and $\qquad |\mathbf{b}| = b = \sqrt{(2)^2 + (1)^2 + (-1)^2}$

$$= 2.45$$

Thus from Equation [1] $\sin \theta = \dfrac{7.87}{(3.74)(2.45)}$

giving $\qquad \theta = 59.2°$

Applications of the Vector Product

(1) The area of the parallelogram shown in Fig. 21.54 is given by:

$$A = |\overrightarrow{AD} \times \overrightarrow{AB}|$$

Fig. **21.54**

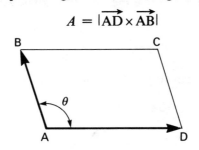

(2) The area of the triangle shown in Fig. 21.55 is given by:

$$A = \tfrac{1}{2}|\overrightarrow{AC} \times \overrightarrow{AB}|$$

Fig. **21.55**

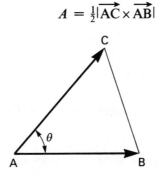

(3) The velocity of a point P on a rotating rigid body is given by the vector product $\boldsymbol{\omega} \times \mathbf{r}$ (Fig. 21.56). In the diagram the z-axis is the axis of rotation and the angular velocity $\boldsymbol{\omega}(0, 0, \omega_z)$ is a vector along this axis. The position vector of P (in the yz plane) is $\mathbf{r}(0, r_y, r_z)$. The velocity of P is given by

$$\mathbf{v} = \boldsymbol{\omega} \times \mathbf{r} = \begin{vmatrix} \mathbf{i} & \mathbf{j} & \mathbf{k} \\ 0 & 0 & \omega_z \\ 0 & r_y & r_z \end{vmatrix} = -r_y\omega_z\mathbf{i}$$

Fig. **21.56**

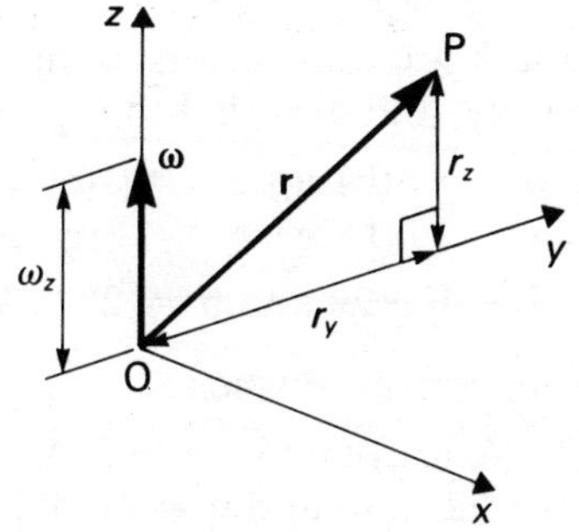

Example 21.23

A triangle ABC has its vertices at the points A(3, 6, 3), B(3, 0, 9) and C(−3, 6, −3). Calculate the area of triangle ABC.

Fig. **21.57**

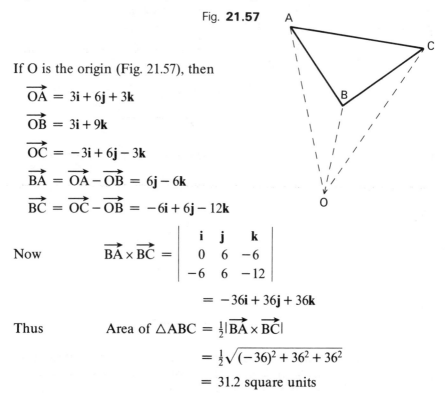

If O is the origin (Fig. 21.57), then

$$\overrightarrow{OA} = 3\mathbf{i} + 6\mathbf{j} + 3\mathbf{k}$$

$$\overrightarrow{OB} = 3\mathbf{i} + 9\mathbf{k}$$

$$\overrightarrow{OC} = -3\mathbf{i} + 6\mathbf{j} - 3\mathbf{k}$$

$$\overrightarrow{BA} = \overrightarrow{OA} - \overrightarrow{OB} = 6\mathbf{j} - 6\mathbf{k}$$

$$\overrightarrow{BC} = \overrightarrow{OC} - \overrightarrow{OB} = -6\mathbf{i} + 6\mathbf{j} - 12\mathbf{k}$$

Now
$$\overrightarrow{BA} \times \overrightarrow{BC} = \begin{vmatrix} \mathbf{i} & \mathbf{j} & \mathbf{k} \\ 0 & 6 & -6 \\ -6 & 6 & -12 \end{vmatrix}$$

$$= -36\mathbf{i} + 36\mathbf{j} + 36\mathbf{k}$$

Thus
Area of $\triangle ABC = \tfrac{1}{2}|\overrightarrow{BA} \times \overrightarrow{BC}|$

$$= \tfrac{1}{2}\sqrt{(-36)^2 + 36^2 + 36^2}$$

$$= 31.2 \text{ square units}$$

EXERCISE

21·5

1 Calculate the magnitude of the vector product of the following vectors:
(a) $|\mathbf{a}| = 2, |\mathbf{b}| = 3, \theta = 60°$
(b) $|\mathbf{a}| = 0.5, |\mathbf{b}| = 4, \theta = 0°$

2 If $\mathbf{i}, \mathbf{j}$ and $\mathbf{k}$ are unit vectors along the x-, y- and z-axes respectively, and $\mathbf{a} = 4\mathbf{i}$, $\mathbf{b} = 8\mathbf{j}$ and $\mathbf{c} = -6\mathbf{k}$, calculate:
(a) $\mathbf{a} \times \mathbf{b}$ (b) $\mathbf{a} \times \mathbf{c}$ (c) $\mathbf{c} \times \mathbf{a}$ (d) $\mathbf{b} \times \mathbf{c}$
(e) $\mathbf{a} \times \mathbf{a}$ (f) $\mathbf{c} \times \mathbf{b}$ (g) $\mathbf{b} \times \mathbf{b}$

3 If $\mathbf{a} = 2\mathbf{i} + 4\mathbf{j} - 2\mathbf{k}$ and $\mathbf{b} = 2\mathbf{j} + 2\mathbf{k}$ find $\mathbf{a} \times \mathbf{b}$.

4 Calculate $\mathbf{c} = \mathbf{a} \times \mathbf{b}$ when:
(a) $\mathbf{a} = (4, 6, 2)$ and $\mathbf{b} = (-2, 4, 8)$
(b) $\mathbf{a} = (-2, 1, 0)$ and $\mathbf{b} = (1, 4, 3)$

5 If $\mathbf{a} = \mathbf{i} + \mathbf{j}$ and $\mathbf{b} = 2\mathbf{i} + \mathbf{k}$, find the angle between $\mathbf{a}$ and $\mathbf{b}$.

6 Which of the following pairs of vectors are perpendicular?
   (a) $\mathbf{a} = (2, -3, 1)$, $\mathbf{b} = (-1, 4, 2)$
   (b) $\mathbf{a} = (-3, 6, -15)$, $\mathbf{b} = (-24, 3, 6)$
   (c) $\mathbf{a} = (8, 4, 4)$, $\mathbf{b} = (2, -8, 4)$

7 The vectors $4\mathbf{i} + 6\mathbf{j} - 2\mathbf{k}$ and $2\mathbf{i} + 4\mathbf{j} + 2\mathbf{k}$ represent two sides of a triangle. Calculate the area of the triangle.

8 The triangle ABC has vertices at the points A(0, 0, 1), B(1, 0, 1) and C(2, 1, 3). Find the area of $\triangle$ABC.

Answers

1·1

1. (a) 15.80   (b) 1.968
   (c) 1.094   (d) 0.0314
   (e) 0.581   (f) 0.925
2. (a) 852.8   (b) 39.28
   (c) 2.179
3. 0.003 573   4. 0.011 82
5. 0.5896   6. 25.7
7. 131.8   8. 0.741
9. 44.24   10. 0.005 49
11. 0.769

1·2

1. $a = 3, n = 0.5$
2. $n = 4.05$, for $I = 20$ read $I = 35$
3. $t = 0.3m^{1.5}$
4. $k = 100, n = -1.2$
5. $a = 245, b = 33$
6. $\mu = 0.5, k = 5$
7. $I = 0.02, T = 0.2$
8. $k = 23.3, c = 2.99$
9. $V = 100, t = 0.0025$

2·1

1. (a) $(8.60, 54.5°)$ (b) $(3.61, 123.7°)$
   (c) $(3.61, 303.7°)$ (d) $(5, 233.1°)$
2. (a) $(1.64, 1.15)$ (b) $(-1.81, 2.39)$
   (c) $(-0.75, -1.30)$ (d) $(0.401, -0.446)$
   (e) $(2.15, -0.824)$ (f) $(3.21, -3.83)$

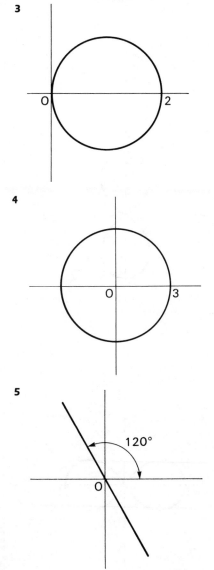

6

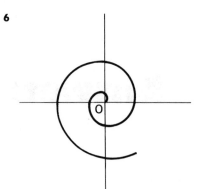

7

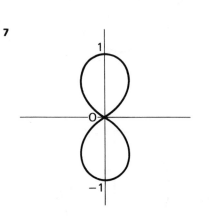

8

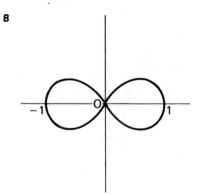

9

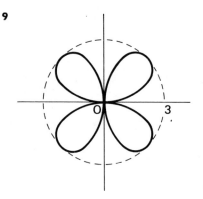

10

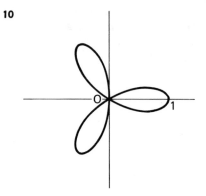

11

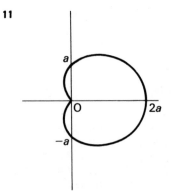

3·1

1 33, 60
2 23rd
3 (a) 27   (b) 203
4 3.75, 4.20, 4.65, 5.10, 5.55, 6.00
5 $-10, 2, 14$ or $6, -8\frac{2}{3}, -23\frac{1}{3}$
6 Nine
7 £274 750
8 88.3 hours
9 (a) 720 m   (b) 135 m
    (c) 24.5 seconds
10 40, 110, 179, 249, 318

3·2

1 1.949
2 -2187
3 $\frac{3}{8}, \frac{1}{8}, \frac{1}{24}$
4 (a) 118.1   (b) 120
5 3, 6, 12, 24, 48, 96
6 6, 11, 22, 42, 81, 156 rev/min
7 47, 78, 129, 213, 353, 584 rev/min
8 (a) 255, 673, 1092, 1510, 1928, 2346, 2765,
         3183 rev/min
    (b) 255, 366, 525, 752, 1079, 1547, 2219,
         3183 rev/min
9 (a) £31 789   (b) £120 000
10 (a) 15.1   (b) 333
11 (a) 4877   (b) 1706

4·1

1 9.17 m, 24.20°   2 257.3 mm, 97.03°
3 111.5 mm, 77.10°   4 26.17 mm
5 144.5 mm   6 4.78 m, 4.09 m
7 13.2 m^2 8 9330 mm^2
9 4105 mm^2 10 576 mm^2
11 50.97°   12 1233 m
13 190.8 m

4·2

1 (a) 42.72 mm   (b) 854.4 mm^2
    (c) 54.18°
2 (a) 78.5 mm   (b) 20.3 mm
    (c) 12.23°
3 (a) 34.92°   (b) 31.30°
4 180 000 mm^2
5 (a) 3.37 m   (b) 25.52°
    (c) 7.83 m

6 86.0 mm, 94.9 mm, 24.93°
7 (a) 50.77°   (b) 104.50°
8 28.73°

5·1

1 $1 + 5z + 10z^2 + 10z^3 + 5z^4 + z^5$
2 $p^6 + 6p^5q + 15p^4q^2 + 20p^3q^3 + 15p^2q^4$
    $+ 6pq^5 + q^6$
3 $x^4 - 12x^3y + 54x^2y^2 - 108xy^3 + 81y^4$
4 $32p^5 - 80p^4q + 80p^3q^2 - 40p^2q^3$
    $+ 10pq^4 - q^5$
5 $128x^7 + 448x^6y + 672x^5y^2 + 560x^4y^3$
    $+ 280x^3y^4 + 84x^2y^5 + 14xy^6 + y^7$
6 $x^3 + 3x + \dfrac{3}{x} + \dfrac{1}{x^3}$
7 $1 + 12x + 66x^2 + 220x^3 + \ldots$
8 $1 - 28x + 364x^2 - 2912x^3 + \ldots$
9 $p^{16} + 16p^{15}q + 120p^{14}q^2 + 560p^{13}q^3 + \ldots$
10 $1 + 30y + 405y^2 + 3240y^3 + \ldots$
11 $x^{18} - 27x^{16}y + 324x^{14}y^2 - 2268x^{12}y^3 + \ldots$
12 $x^{22} + 11x^{18} + 55x^{14} + 165x^{10} + \ldots$

5·2

1 $1 + 6x + 15x^2 + 20x^3 + \ldots$
2 $1 + 18x + 144x^2 + 672x^3 + \ldots$
3 $1 - 4x + 10x^2 - 20x^3 + \ldots$
4 $1 + \dfrac{x}{2} + \dfrac{3x^2}{8} + \dfrac{5x^3}{16} + \ldots$
5 $1 - x + x^2 - x^3 + \ldots$
6 $1 - 6x + 24x^2 - 80x^3 + \ldots$
7 $1 + 3x + 6x^2 + 10x^3 + \ldots$
8 $1 + \dfrac{x^2}{3} - \dfrac{x^4}{9} + \dfrac{5}{81}x^6 - \ldots$
9 $1 + \dfrac{3x}{2} + \dfrac{27x^2}{8} + \dfrac{135x^3}{16} + \ldots$
10 $1 - \dfrac{x}{2} - \dfrac{x^2}{24} - \dfrac{5x^3}{432} - \ldots$

5·3

2 1% too large
3 11% too small
4 3% decrease
5 18% increase
6 $\dfrac{3}{\sqrt{2}}\left(1 - \dfrac{x}{4} + \dfrac{3x^2}{32}\right)$
8 2.06% decrease
9 1.5%
11 2% too small

6·1

1 (a) 4.48   (b) 22.2
   (c) 0.135   (d) 0.0136
   (e) 1.22   (f) 1.70
   (g) 0.497

2 (a) $1 + 3x + \dfrac{(3x)^2}{2!} + \dfrac{(3x)^3}{3!} + \ldots$

   (b) $1 + 0.5x + \dfrac{(0.5x)^2}{2!} + \dfrac{(0.5x)^3}{3!} + \ldots$

   (c) $1 - 1.3x + \dfrac{(1.3x)^2}{2!} - \dfrac{(1.3x)^3}{3!} + \ldots$

   (d) $1 - 0.3x + \dfrac{(0.3x)^2}{2!} - \dfrac{(0.3x)^3}{3!} + \ldots$

3 (a) 7.051   (b) 1.020

4 $1 + \dfrac{x^2}{2!} + \dfrac{x^4}{4!} + \dfrac{x^6}{6!} + \ldots$

7·1

No answers

7·2

4 (a) $\sin\left(\omega t + \dfrac{\pi}{2}\right)$

   (b) $\sin(\omega t - \pi)$
   (c) $\sin \omega t$

   (d) $\sin\left(\omega t - \dfrac{\pi}{6}\right)$

6 3.61, 33.7°
7 $v_r = 7.7 \sin(\theta - 19°)$
8 $i_r = 6.6 \sin(\theta + 12°)$
9 $v_r = 51 \sin(\theta - 11°)$

8·1

1 (a) $-\sin x$ (b) $-\cos x$
   (c) $\cos x$ (d) $\sin x$
   (e) $\tan x$

2 0.259

4 (a) $\frac{16}{65}$ (b) $\frac{33}{65}$ (c) $\frac{56}{33}$

   (d) $\frac{24}{25}$ (e) $\frac{119}{169}$ (f) $\frac{24}{7}$

6 46.8°
7 (b) 0.960, 0.280, −3.429
8 10.9°
9 18.4°
10 (a) −0.551   (b) −0.551

8·2

1 3.61 sin $(\theta + 33.7°)$
2 7.071 sin $(\omega t + 1.429)$
3 8.602 sin $(\omega t - 0.951)$
4 8.602, 2.521 radians
5 223.6 sin $(300t + 0.464)$, 223.6
6 7.74 sin $(\theta - 18.8°)$
7 6.57 sin $(\theta + 12.4°)$
8 51.2 sin $(\theta - 10.6°)$

9·1

1 (a) $18 + j10$ (b) $-j$
   (c) $8 - j10$
2 (a) $-1 + j3$ (b) $-4 - j3$
   (c) $10 - j3$
3 (a) $-9 + j21$ (b) $-36 - j32$
   (c) $-9 + j40$ (d) 34
   (e) $-21 - j20$ (f) $18 - j30$
   (g) $0.069 - j0.172$ (h) $-0.724 + j0.690$
   (i) $-0.138 - j0.655$ (j) $0.644 + j0.616$
   (k) $3.5 - j0.5$ (l) $0.2 + j0.6$
4 (a) $1, j0.5$ (b) $0, j5$
   (c) $-3, j3$
5 (a) $-1 \pm j$ (b) $\pm j3$
6 (a) $0.634 - j0.293$
7 (a) $3 + j4$ (b) $0.4 + j0.533$
   (c) $0.692 + j2.538$

9·2

2 Mod 5, Arg 53.13°;
   Mod 5, Arg −36.87°
3 3.61, 146.32°
4 4.47, −153.43°
5 (a) $5\,\underline{/36.87°}$ (b) $5\,\underline{/-53.13°}$
   (c) $4.24\,\underline{/135°}$ (d) $2.24\,\underline{/-153.43}$
   (e) $4\,\underline{/90°}$ (f) $3.5\,\underline{/-90°}$
6 (a) $2.12 + j2.12$ (b) $-4.49 + j2.19$
   (c) $4.32 - j1.57$ (d) $-1.60 - j2.77$
7 (a) $56\,\underline{/70°}$ (b) $10\,\underline{/-50°}$
   (c) $15\,\underline{/90°}$ (d) $21\,\underline{/-90°}$
8 (a) $2.67\,\underline{/-30°}$ (b) $2\,\underline{/-60°}$
   (c) $0.6\,\underline{/-9°}$ (d) $2.83\,\underline{/44.65°}$
9 (a) $30\,\underline{/13°}$ (b) $-2.93 + j6.90$
10 (a) $0.277\,\underline{/56.32°}$ (b) $13\,\underline{/-112.63°}$
11 (a) $2.83\,\underline{/19°}$ (b) $1.674 + j0.896$
12 (a) $0.172\,\underline{/-59.03°}$ (b) $0.0427 - j0.0385$

13 (a) $r = 4.5, X_L = 2.2$
(b) $R = 23, X_C = 35$
(c) $R = 27.2, X_L = 11.7$
(d) $R = 6.85, X_C = 1.46$
14 (a) $14.62°$ (b) 345 watts

10·1

1 $15x^2 + 14x - 1$
2 $3.5t^{-0.5} - 1.8t^{-0.7}$
3 1.25 **4** 1   **5** $5, -3$

10·2

1 $6(3x + 1)$ **2** $-15(2 - 5x)^2$
3 $-2(1 - 4x)^{-1/2}$ **4** $-7.5(2 - 5x)^{-1/2}$
5 $-4(4x^2 + 3)^{-2}$ **6** $3 \cos (3x + 4)$
7 $5 \sin (2 - 5x)$ **8** $8 \sin 4x \cos 4x$
9 $21 (\sin 7x)/\cos^4 7x$

10 $2 \cos \left(2x + \dfrac{\pi}{2} \right)$

11 $-3(\sin x)\cos^2 x$

12 $-\dfrac{\cos x}{\sin^2 x}$

13 $\dfrac{1}{x}$ **14** $-\dfrac{9}{x}$

15 $\dfrac{1}{2(2x - 7)}$ **16** $-\dfrac{1}{e^x}$

17 $6e^{(3x + 4)}$ **18** $8e^{(8x - 2)}$
19 $\frac{2}{3}(1 - 2t)^{-4/3}$
20 $\frac{3}{4} \cos (\frac{3}{4}\theta - \pi)$
21 $[-\sin (\pi - \phi)]/\cos^2(\pi - \phi)$

22 $-\dfrac{1}{2x}$ **23** $Bke^{(kt - b)}$

24 $-\dfrac{1}{3}e^{(1 - x)/3}$

10·3

1 (a) $\sin x + x \cos x$
(b) $e^x (\tan x + \sec^2 x)$
(c) $1 + \log_e x$
2 $\cos^2 t - \sin^2 t$
3 $2(\tan \theta) \cos 2\theta + (\sec^2 \theta) \sin 2\theta$
4 $e^{4m}(4 \cos 3m - 3 \sin 3m)$
5 $3x(1 + 2 \log_e x)$
6 $6e^{3t}(3t^2 + 2t - 3)$
7 $1 - 3z + (1 - 6z)\log_e z$

8 (a) $\dfrac{1}{(1 - x)^2}$ (b) $\dfrac{1 - 2 \log_e x}{x^3}$

(c) $\dfrac{e^x(\sin 2x - 2 \cos 2x)}{\sin^2 2x}$

9 $\dfrac{11}{(3 - 4z)^2}$

10 $-\dfrac{2(\sin 2t + \cos 2t)}{e^{2t}}$

11 $-\text{cosec}^2\theta$

10·4

1 21.59   **2** -3.77
3 1.005   **4** -0.3573
5 0   **6** 26.15
7 -5.882 **8** -4.050
9 0.5   **10** 0.5266

11·1

1 $18x, 54$ **2** $-187, 254$

3 29.3   **4** $-\dfrac{1}{x^2}, -0.277$

5 $15, 0$ **6** -89.0
7 -0.115 **8** 7.52
9 $-4.6, -21.2$
10 222   **11** 2.94

11·2

1 42 m/s   **2** -6 m/s^2
3 (a) 6 m/s   (b) 2.41 or -0.41 s
(c) 6 m/s^2 (d) 1 s
4 -0.074 m/s, 0.074 m/s^2
5 10 m/s, 30 m/s
6 3.46 m/s
7 (a) 4 rad/s   (b) 36 rad/s^2
(c) 0 s or 1 s
8 (a) -2.97 rad/s   (b) 0.280 s
(c) -8.98 rad/s^2 (d) 1.57 s
9 62.5 kJ

12·1

1 (a) 11 (max), -16 (min)
(b) 4 (max), 0 (min)
(c) 0 (min), 32 (max)
2 (a) 54   (b) $x = 2.5$
(c) $x = -2$

3 $(3, -15), (-1, 17)$

4 (a) -2 (b) 1

   (c) 9

5 (a) 12   (b) 12.48

6 15 mm   **7** 10 m

8 4   **9** 108 000 mm³

10 Radius = Height = 4.57 m

11 405 mm

12 Diameter = 28.9 mm

   Height 14.4 mm

13 5.76 m   **14** 2.15

15 84.8 mm   **16** $\cos^2\alpha$

13·1

1 $\dfrac{2}{(x-3)} + \dfrac{3}{(x+3)}$

2 $\dfrac{2}{(x-2)} - \dfrac{1}{(x-1)}$

3 $\dfrac{4}{(x+1)} - \dfrac{3}{(2x-1)}$

4 $\dfrac{6}{(x+1)} - \dfrac{5}{(x+2)}$

5 $\dfrac{2}{(x+1)} + \dfrac{3}{(x+2)} + \dfrac{4}{(x+3)}$

6 $\dfrac{1}{(2x-1)} + \dfrac{2}{(x+1)} - \dfrac{1}{(x-1)}$

7 $\dfrac{3}{2x} + \dfrac{1}{(1-x)} - \dfrac{2}{(2-3x)}$

8 $\dfrac{1}{(1+x)} + \dfrac{1}{(x-1)} - \dfrac{4}{(2x+1)}$

9 $\dfrac{1}{(2-x)} + \dfrac{2}{(2-x)^2} - \dfrac{1}{(1+x)}$

10 $\dfrac{1}{(2x+3)} + \dfrac{1}{(x-1)^2} - \dfrac{2}{(x-1)}$

11 $\dfrac{1}{(x+1)} + \dfrac{2}{(x-1)^2}$

12 $\dfrac{2}{(1+x)} - \dfrac{2}{(1+x)^2} - \dfrac{2}{x}$

13 $\dfrac{2x+1}{(x^2+1)} + \dfrac{1}{(x+1)}$

14 $\dfrac{3x+5}{(2+x^2)} - \dfrac{1}{(x-1)}$

15 $\dfrac{1+x}{(2-3x^2)} + \dfrac{3}{(4+x)}$

16 $\dfrac{3x+2}{(x^2+4)} + \dfrac{3}{(x+2)}$

17 $1 + \dfrac{1}{(x-1)} + \dfrac{2}{(x+2)}$

18 $1 + \dfrac{2}{(2x-1)} - \dfrac{1}{(x-1)}$

19 $1 - \dfrac{2}{(x-1)} - \dfrac{1}{(x+1)}$

20 $3 - \dfrac{1}{(x+1)} + \dfrac{1}{(x-2)}$

14·1

1 $\frac{5}{3}x^3 + x^2 + \dfrac{4}{x} + c$

2 $\frac{2}{3}x^{3/2} + 2x^{1/2} + c$

3 $-3\left(\cos\dfrac{x}{3}\right) + c$

4 $\frac{5}{3}(\sin 3\theta) + c$

5 $\phi - \frac{3}{2}(\cos\frac{2}{3}\phi) + c$

6 $2\left(\sin\dfrac{\theta}{2}\right) + \frac{2}{3}\left(\cos\dfrac{3\theta}{2}\right) + c$

7 $t^2 - \frac{1}{2}(\cos 2t) + c$

8 $\frac{1}{3}e^{3x} + c$

9 $-2e^{-0.5u} + c$

10 $\frac{3}{2}e^{2t} - 2e^t + c$

11 $-2e^{-x/2} + \frac{2}{3}e^{3x/2} + c$

12 $(\tan x) + (\log_e x) + c$

13 7.75   **14** 4.67

15 0.561   **16** 0

17 0.521   **18** 0

19 0.586   **20** 1.12

21 1.33   **22** 1.72

23 0.0585   **24** 0.253

25 4.15   **26** 51.1

27 3.49   **28** 2.18

29 0.811   **30** 0.732

14·2

1 136   **2** 87

3 5.167   **4** 0.667

5 3.75   **6** 0

7 4   **8** 2

9 6.39   **10** 5.21

14·3

1 (a) 0.637 V   (b) 0
2 (a) 3.82   (b) 0
3 1 volt   4 0 volt
5 5 volt   6 1.67 volt
7 2.25 volt
8 (a) 0.707 V   (b) 0.707 V
9 (a) 4.24   (b) 4.24
10 1.42, 2, 5.77, 2.36, 2.45 volts

14·4

1 $\frac{1}{8}(2x+1)^4 + c$
2 $\frac{2}{3}(x+3)^{3/2} + c$
3 $-\frac{1}{2}\log_e(1-2x) + c$
4 $-\frac{1}{8}(3-x^2)^4 + c$
5 $-\frac{1}{2}\cos(2\theta - 1) + c$
6 $-\frac{1}{2}e^{(1-2x)} + c$
7 0.693   8 27.7
9 0.0855   10 $\frac{1}{3}$
11 4024   12 0.3574
13 0.25   14 0.333
15 2.054   16 $\frac{2}{3}$
17 0.3133   18 -0.2152
19 1.228   20 $\frac{4}{3}$
21 0.3466

14·5

1 $\arcsin x + c$ 2 1.107
3 $\frac{1}{2}\arctan \frac{x}{2} + c$ 4 3.861
5 0.1745   6 $\frac{1}{2}\arcsin 2x + c$
7 0.0522   8 0.206
9 0.0432
10 $\frac{1}{\sqrt{2}} \log_e \tan \frac{1}{2}\left(\theta + \frac{\pi}{4}\right) + c$
11 0.861   12 0.7854

14·6

1 $2\log_e(x-3) + 3\log_e(x+3) + c$
2 0.981
3 $2\log_e(x+1) + 3\log_e(x+2) + 4\log_e(x+3) + c$
4 -1.005
5 $\frac{3}{2}\log_e x - \log_e(1-x) + \frac{2}{3}\log_e(2-3x) + c$
6 0.503

7 $\frac{2}{(2-x)} - \log_e(2-x) - \log_e(1+x) + c$
8 -2.901
9 $\frac{1}{(1+x)} + 2\log_e(1+x) - 2\log_e x + c$
10 4.938
11 1.890
12 3.910

14·7

1 $xe^x - e^x + c$
2 $\frac{3}{4}x^2(2\log_e 5x - 1) + c$
3 $\frac{2}{3}xe^{3x} - \frac{2}{9}e^{3x} + c$
4 0.571   5 2.142
6 0.386   7 0.265
8 2   9 0.494
10 $(x^2 - 2)\sin x + 2x\cos x$
11 $\frac{2}{9}x \sin 3x + \left(\frac{2}{27} - \frac{x^2}{3}\right)\cos 3x$
12 0.0289
13 $\frac{1}{2}e^t(\sin t + \cos t)$
14 2.079
15 $\frac{e^{ax}}{a^2 + b^2}(b \sin bx + a \cos bx)$

15·1

1 (a) 6.89   (b) 7.12
   (c) 7.00
2 (a) 0.851   (b) 0.846
   (c) 0.848
3 (a) 0.386   (b) 0.387
   (c) 0.386
4 (a) 0.368   (b) 0.365
   (c) 0.366
5 (a) 0.507   (b) 0.495
   (c) 0.499
6 (a) 0.0176   (b) 0.0169
   (c) 0.0171
7 (a) 0.647   (b) 0.647
   (c) 0.647
8 Both 10 and 12 intervals give 2.42

16·1

1 $y = \frac{1}{2}x^2 + 1$
2 $y = x^3 - 128$
3 $y = \frac{1}{3}x^3 - \frac{5}{2}x^2 + k, y = \frac{1}{3}x^3 - \frac{5}{2}x^2 + 18.7$
4 $y = 1.5x^4 + 1.67x^3 + 7x - 49.4$
5 $y = x^2$

16·2

1 $y = Ae^{3x}, y = 2e^{3x}, 807$
2 $s = 12.25e^{-0.560t}, 3.24$
3 $I = 10e^{-33.3t}, 5.14\,A$
4 20 years
5 $Q = 0.001\,5e^{-41.7t},$
   0.000 652 coulomb,
   0.009 72 seconds

16·3

1 First order, first degree
2 Third order, first degree
3 Second order, first degree
4 First order, second degree
5 Fourth order, first degree
6 Second order, first degree
7 Second order, first degree
8 Third order, second degree
9 Fourth order, first degree
10 Second order, first degree

16·4

1 (a) $\dfrac{d^2y}{dx^2} = 2$ (b) $\dfrac{dy}{dx} = x + \dfrac{y}{x}$

2 $x\dfrac{dy}{dx} = 2x - y$

3 $\dfrac{x^2}{2}\left(\dfrac{d^2y}{dx^2}\right) - x\left(\dfrac{dy}{dx}\right) + y = 0$

4 $x^2\left(\dfrac{d^2y}{dx^2}\right) - 3x\left(\dfrac{dy}{dx}\right) + 3y = 0$

5 $\dfrac{d^2y}{dt^2} = -\omega^2 y$

6 $\dfrac{d^2\theta}{dt^2} = -\theta$

7 $s = ut - \tfrac{1}{2}gt^2$

8 1.17×10^{-4}

16·5

1 $y = x^3 + 3x^2 + A$

2 $y = \dfrac{x^3}{3} + 6\log_e x + A$

3 $y = \sqrt{2x + A}$

4 $y = \sqrt{2x + 0.5x^2 + A}$

5 $y = \tfrac{1}{2}\sin 2x - \tfrac{1}{3}\cos 3x + A$
6 $y = \tfrac{1}{2}e^{2x} + 6x + A$
7 $y = \sqrt{x^2 + A}$
8 $y = Ae^{x^2/2}$
9 $y = Ax$
10 $y = \sqrt[3]{1.5x^2 + 3x + A}$
11 $\theta = \arcsin(0.5t^2 + A)$
12 $y = \sqrt[3]{1.5x^2 + 0.75x^4 + A}$
13 $y = \sqrt{2.755 - 2\cos x}$
14 $y = -3e^{-x} + 5$
15 $y = \sqrt{5.37 + 2\tan x}$
16 $y = x + \log_e x - 2.693$
17 $y = \log_e(\cos x) + 0.231$
18 $y = \log_e(1.718 + e^x)$
19 $v = \sqrt{6t^2 + 14t + 1}$

20 $v = \dfrac{1}{3 - \arctan u}$

17·1

1 $\bar{x} = 38.7\,mm, \bar{y} = 41.8\,mm$
2 $\bar{x} = 68.3\,mm, \bar{y} = 73.8\,mm$
3 $\bar{x} = 19.1\,mm, \bar{y} = 28.3\,mm$
4 $\bar{x} = 50.0\,mm, \bar{y} = 44.6\,mm$
5 $\bar{y} = 42.3\,mm$
6 $\bar{x} = 86.7\,mm, \bar{y} = 60.7\,mm$
7 $\bar{y} = 29.7\,mm$
8 $\bar{y} = 10.6\,mm$

17·2

1 57.4   2 402
3 0.0761   4 171
5 16.8   6 0.105
7 (a) $262\,000\,mm^3$ (b) $58\,700\,mm^3$
8 14.7 litres
9 $23\,100\,mm^3$

17·3

1 $\tfrac{1}{2}L$
2 $\tfrac{3}{4}H$ from apex
3 0.911 m from small flat surface
4 1.33 m from large flat surface
5 0.688 from flat surface

18·1

1 $I_{XX} = 856 \text{ cm}^4, I_{YY} = 1184 \text{ cm}^4$
2 $I_{XX} = 1406 \text{ cm}^4, I_{YY} = 354 \text{ cm}^4$
3 $I_{XX} = 603 \text{ cm}^4, I_{YY} = 363 \text{ cm}^4$
4 $I_{XX} = 459 \text{ cm}^4, I_{YY} = 419 \text{ cm}^4$

18·2

1 $I_{XX} = 493 \text{ cm}^4, I_{YY} = 173 \text{ cm}^4$
2 $I_{XX} = 523 \text{ cm}^4, I_{YY} = 1100 \text{ cm}^4$
3 $I_{XX} = 136 \text{ cm}^4, I_{YY} = 60 \text{ cm}^4$
4 $I_{XX} = 375 \text{ cm}^4, I_{YY} = 266 \text{ cm}^4$
5 $I_{XX} = 133 \text{ cm}^4, I_{YY} = 227 \text{ cm}^4$
6 $I_{XX} = 274 \text{ cm}^4, I_{YY} = 219 \text{ cm}^4$

18·3

1 (a) 258 cm^4　　(b) 517 cm^4
　(c) 2520 cm^4
2 (a) $I_{XX} = I_{YY} = 168 \text{ cm}^4, J = 336 \text{ cm}^4$
　(b) $k_{XX} = 2.34 \text{ cm}$
3 (a) 1600 cm^4　　(b) 1820 cm^4
　(c) 909 cm^4

19·1

1 0.0133 kg m^2　　**2** 104 mm
3 21.4 kg m^2
4 (a) 8390 kg m^2　　(b) $32\,300 \text{ kg m}^2$
5 (a) 288 kg m^2　　(b) 1050 kg m^2
6 $17\,700 \text{ kg m}^2$　　**7** 7.28 kg m^2

20·1

1 (a) 2×2　　(b) 2×1
　(c) 3×3　　(d) 2×4
2 (a) 9　　(b) 4
　(c) n^2
3 (a) $\begin{pmatrix} 1 & 3 \\ 2 & 4 \end{pmatrix}$　　(b) $(5 \quad -6)$
　(c) $\begin{pmatrix} a & 2 & x \\ b & 3 & -6 \\ 4 & 5 & 0 \end{pmatrix}$　(d) $\begin{pmatrix} 1 & 6 \\ -2 & 2 \\ -3 & 0 \\ -4 & -1 \end{pmatrix}$
4 (a) $\begin{pmatrix} 0 & 0 \\ 9 & 2 \end{pmatrix}$　　(b) $\begin{pmatrix} 4 & 2 \\ -3 & 2 \end{pmatrix}$
　(c) $\begin{pmatrix} \frac{5}{6} & \frac{1}{2} \\ \frac{5}{6} & 1 \end{pmatrix}$

5 $a = -8, b = 5, c = 1$
6 $\begin{pmatrix} \frac{1}{3} & \frac{1}{20} \\ \frac{1}{30} & \frac{1}{18} \end{pmatrix}$　　**7** $\begin{pmatrix} 5 & 8 \\ 12 & -2 \end{pmatrix}$
8 $\begin{pmatrix} 0 \\ 5 \end{pmatrix}$

20·2

1 (a) $\begin{pmatrix} 6 & 0 \\ -4 & 2 \end{pmatrix}$　　(b) $\begin{pmatrix} -12 & 3 \\ 9 & -6 \end{pmatrix}$
　(c) $\begin{pmatrix} -6 & 3 \\ 5 & -4 \end{pmatrix}$　　(d) $\begin{pmatrix} 18 & -3 \\ -13 & 8 \end{pmatrix}$
2 (a) $\begin{pmatrix} 14 & 0 \\ 8 & -2 \end{pmatrix}$　　(b) $\begin{pmatrix} 2 & 1 \\ 3 & 1 \end{pmatrix}$
　(c) $\begin{pmatrix} 5 & 11 \\ 10 & 22 \end{pmatrix}$　　(d) $\begin{pmatrix} a & b \\ c & d \end{pmatrix}$
　(e) $\begin{pmatrix} ka & kb \\ kc & kd \end{pmatrix}$
3 (a) $\begin{pmatrix} 7 & 10 \\ 15 & 22 \end{pmatrix}$　　(b) $\begin{pmatrix} 3 & -5 \\ 5 & 8 \end{pmatrix}$
　(c) $\begin{pmatrix} 8 & 10 \\ 20 & 18 \end{pmatrix}$　　(d) $\begin{pmatrix} 18 & 15 \\ 40 & 48 \end{pmatrix}$
　(e) $\begin{pmatrix} 13 & 10 \\ 40 & 53 \end{pmatrix}$

20·3

1 (a) 24　　(b) -6
　(c) 14
2 (a) $x = 1, y = 2$　　(b) $x = 4, y = 3$
　(c) $x = 0.5, y = 0.75$

20·4

1 $\frac{1}{3}\begin{pmatrix} 4 & -5 \\ -1 & 2 \end{pmatrix}$　　**2** $\begin{pmatrix} 3 & -5 \\ -1 & 2 \end{pmatrix}$
3 $\frac{1}{4}\begin{pmatrix} 2 & -2 \\ -1 & 3 \end{pmatrix}$　　**4** No inverse
5 $\frac{1}{320}\begin{pmatrix} 4 & -24 \\ -24 & 224 \end{pmatrix}$
6 No inverse　　**7** $\frac{1}{7}\begin{pmatrix} 1 & -3 \\ 1 & 2 \end{pmatrix}$
8 $\frac{1}{13}\begin{pmatrix} 5 & 3 \\ -1 & 2 \end{pmatrix}$　　**9** $\begin{pmatrix} 1 & -1 \\ 0 & 1 \end{pmatrix}$

10 (a) $\dfrac{1}{2}\begin{pmatrix} 2 & 0 \\ -3 & 1 \end{pmatrix}$ (b) $\begin{pmatrix} 2 & -5 \\ -1 & 3 \end{pmatrix}$

(c) $\dfrac{1}{2}\begin{pmatrix} 19 & -5 \\ -11 & 3 \end{pmatrix}$

(d) $\begin{pmatrix} 3 & 5 \\ 11 & 19 \end{pmatrix}$ (e) $\dfrac{1}{2}\begin{pmatrix} 19 & -5 \\ -11 & 3 \end{pmatrix}$

(f) equal

20·5

1 $x = 6, y = -5$ 2 $x = 1, y = 5$

3 $x = 3, y = -1$ 4 $x = 2, y = -3$

5 $x = \dfrac{16}{11}, y = \dfrac{9}{11}$ 6 $x = 2, y = -5$

20·6

1 30	2 21	3 −38
4 78	5 −119	6 −143
7 −28	8 −63	9 −120
10 −48	11 −54	12 0
13 −32	14 −4	15 −18
16 0	17 0	18 0
19 43	20 −640	21 26
22 4	23 186	24 $(a-1)^2$

21·1

1 (b), (d), (e) and (f)

3 (a) $\begin{pmatrix} 2 \\ 6 \end{pmatrix}$ (b) $\begin{pmatrix} 0 \\ 4 \end{pmatrix}$

(c) $\begin{pmatrix} -4 \\ 0 \end{pmatrix}$ (d) $\begin{pmatrix} 5 \\ -5 \end{pmatrix}$

4 (i) 4.24

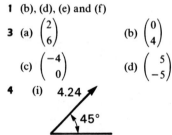

(ii)

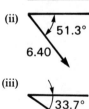

(iii)

(iv)

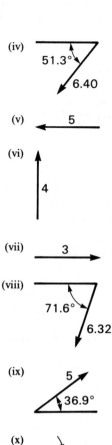

(v) ← 5

(vi) ↑ 4

(vii) → 3

(viii) 71.6° 6.32

(ix) 5 36.9°

(x) 31.0° 5.83

(xi) 36.9° 5

(xii) 7.28 74.1°

5 (a) $\overrightarrow{OA} = 4\mathbf{i} + 6\mathbf{j}$

(b) 7.21 units; 56.3°

56.3°

21·2

1 (b) (i)

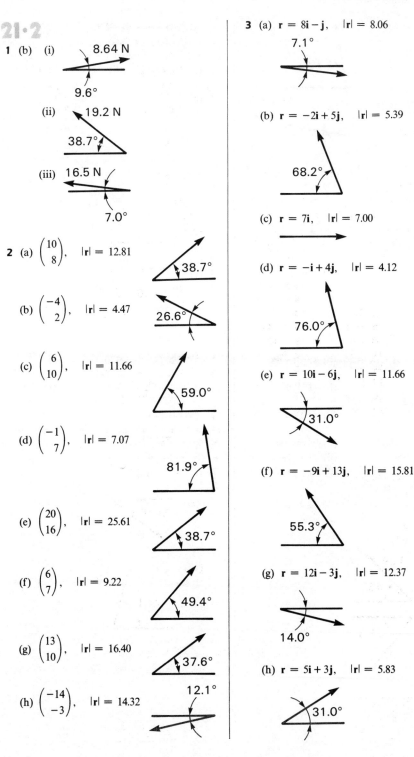

8.64 N

9.6°

(ii) 19.2 N

38.7°

(iii) 16.5 N

7.0°

2 (a) $\begin{pmatrix} 10 \\ 8 \end{pmatrix}$, $|\mathbf{r}| = 12.81$

38.7°

(b) $\begin{pmatrix} -4 \\ 2 \end{pmatrix}$, $|\mathbf{r}| = 4.47$

26.6°

(c) $\begin{pmatrix} 6 \\ 10 \end{pmatrix}$, $|\mathbf{r}| = 11.66$

59.0°

(d) $\begin{pmatrix} -1 \\ 7 \end{pmatrix}$, $|\mathbf{r}| = 7.07$

81.9°

(e) $\begin{pmatrix} 20 \\ 16 \end{pmatrix}$, $|\mathbf{r}| = 25.61$

38.7°

(f) $\begin{pmatrix} 6 \\ 7 \end{pmatrix}$, $|\mathbf{r}| = 9.22$

49.4°

(g) $\begin{pmatrix} 13 \\ 10 \end{pmatrix}$, $|\mathbf{r}| = 16.40$

37.6°

(h) $\begin{pmatrix} -14 \\ -3 \end{pmatrix}$, $|\mathbf{r}| = 14.32$

12.1°

3 (a) $\mathbf{r} = 8\mathbf{i} - \mathbf{j}$, $|\mathbf{r}| = 8.06$

7.1°

(b) $\mathbf{r} = -2\mathbf{i} + 5\mathbf{j}$, $|\mathbf{r}| = 5.39$

68.2°

(c) $\mathbf{r} = 7\mathbf{i}$, $|\mathbf{r}| = 7.00$

(d) $\mathbf{r} = -\mathbf{i} + 4\mathbf{j}$, $|\mathbf{r}| = 4.12$

76.0°

(e) $\mathbf{r} = 10\mathbf{i} - 6\mathbf{j}$, $|\mathbf{r}| = 11.66$

31.0°

(f) $\mathbf{r} = -9\mathbf{i} + 13\mathbf{j}$, $|\mathbf{r}| = 15.81$

55.3°

(g) $\mathbf{r} = 12\mathbf{i} - 3\mathbf{j}$, $|\mathbf{r}| = 12.37$

14.0°

(h) $\mathbf{r} = 5\mathbf{i} + 3\mathbf{j}$, $|\mathbf{r}| = 5.83$

31.0°

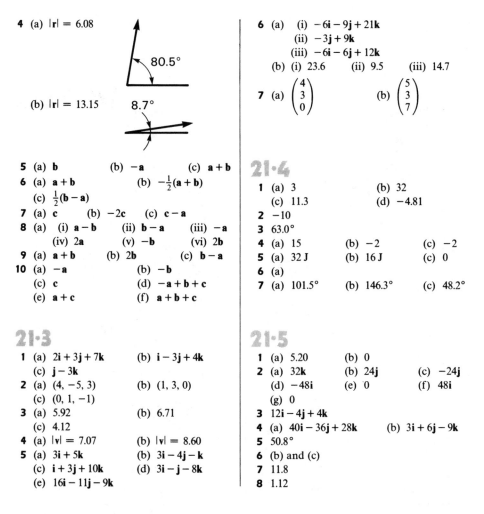

4 (a) |**r**| = 6.08

80.5°

(b) |**r**| = 13.15   8.7°

5 (a) **b**   (b) −**a**   (c) **a** + **b**
6 (a) **a** + **b**   (b) −½(**a** + **b**)
(c) ½(**b** − **a**)
7 (a) **c**   (b) −2**c**   (c) **c** − **a**
8 (a) (i) **a** − **b**   (ii) **b** − **a**   (iii) −**a**
(iv) 2**a**   (v) −**b**   (vi) 2**b**
9 (a) **a** + **b**   (b) 2**b**   (c) **b** − **a**
10 (a) −**a**   (b) −**b**
(c) **c**   (d) −**a** + **b** + **c**
(e) **a** + **c**   (f) **a** + **b** + **c**

21·3

1 (a) 2**i** + 3**j** + 7**k**   (b) **i** − 3**j** + 4**k**
(c) **j** − 3**k**
2 (a) (4, −5, 3)   (b) (1, 3, 0)
(c) (0, 1, −1)
3 (a) 5.92   (b) 6.71
(c) 4.12
4 (a) |**v**| = 7.07   (b) |**v**| = 8.60
5 (a) 3**i** + 5**k**   (b) 3**i** − 4**j** − **k**
(c) **i** + 3**j** + 10**k**   (d) 3**i** − **j** − 8**k**
(e) 16**i** − 11**j** − 9**k**

6 (a) (i) −6**i** − 9**j** + 21**k**
(ii) −3**j** + 9**k**
(iii) −6**i** − 6**j** + 12**k**
(b) (i) 23.6   (ii) 9.5   (iii) 14.7

7 (a) $\begin{pmatrix} 4 \\ 3 \\ 0 \end{pmatrix}$ (b) $\begin{pmatrix} 5 \\ 3 \\ 7 \end{pmatrix}$

21·4

1 (a) 3   (b) 32
(c) 11.3   (d) −4.81
2 −10
3 63.0°
4 (a) 15   (b) −2   (c) −2
5 (a) 32 J   (b) 16 J   (c) 0
6 (a)
7 (a) 101.5°   (b) 146.3°   (c) 48.2°

21·5

1 (a) 5.20   (b) 0
2 (a) 32**k**   (b) 24**j**   (c) −24**j**
(d) −48**i**   (e) 0   (f) 48**i**
(g) 0
3 12**i** − 4**j** + 4**k**
4 (a) 40**i** − 36**j** + 28**k**   (b) 3**i** + 6**j** − 9**k**
5 50.8°
6 (b) and (c)
7 11.8
8 1.12

Index